ENERGY SAVINGS IN
AGRICULTURAL MACHINERY AND MECHANIZATION

This study was carried out under contract at the Centre for Studies and Information of the European Communities upon request of the Directorate-General for Energy, Commission of the European Communities, Brussels, Belgium

ENERGY SAVINGS
IN AGRICULTURAL MACHINERY
AND MECHANIZATION

Edited by

G. PELLIZZI

Institute of Agricultural Engineering, Milan University, Italy

A. GUIDOBONO CAVALCHINI

and

M. LAZZARI

Institute of Agricultural Mechanics, University of Udine, Italy

ELSEVIER APPLIED SCIENCE
LONDON and NEW YORK

ELSEVIER APPLIED SCIENCE PUBLISHERS LTD
Crown House, Linton Road, Barking, Essex IG11 8JU, England

Sole Distributor in the USA and Canada
ELSEVIER SCIENCE PUBLISHING CO., INC.
52 Vanderbilt Avenue, New York, NY 10017, USA

WITH 21 TABLES AND 64 ILLUSTRATIONS

© 1988 ECSC, EEC, EAEC, BRUSSELS AND LUXEMBOURG
Softcover reprint of the hardcover 1st edition 1988

British Library Cataloguing in Publication Data

Energy savings in agricultural machinery and mechanization
1. Agricultural energy savings equipment
I. Pellizzi, G.   II. Cavalchini, A. Guidobono
III. Lazzari, M.
631.3'7

Library of Congress CIP data

Energy savings in agricultural machinery and mechanization.

Bibliography: p.
1. Agriculture—European Economic Community countries
—Energy conservation.   I. Pellizzi, G.   II. Guidobono
Cavalchini, A.   III. Lazzari, M.
TJ163.5.A37E54   1988   631.3   88-11071

ISBN-13: 978-94-010-7108-6   e-ISBN-13: 978-94-009-1365-3
DOI: 10.1007/978-94-009-1365-3

Publication arrangements by Commission of the European Communities, Directorate-General Telecommunications, Information Industries and Innovation, Luxembourg.

EUR 11247

I N T R O D U C T I O N

In the context of the demonstration programme of the Directorate-General
for Energy of the Commission of the European Communities on energy
savings in the agricultural sector, the question of potential energy
savings in agricultural machinery and mechanization is extremely important.

Direct and indirect energy consumption in Community agriculture has a
marked effect on the prices of agricultural products and on market competition , and correct management of agricultural machinery and farming
practices is bound to have a positive effect on the sector's economic
performance.

Since the information available is fragmented and not always directly
comparable, the Directorate-General for Energy has commissioned this
study with the specific aim of quantifying the energy saving potential in
the agricultural sector in a uniform manner so that future energy demonstration projects can be tailored to identified needs.

The study was carried out under contract at the Centre for Studies and
Information of the European Communities by Prof. G. Pellizzi of the Institute of Agricultural Engineering of Milan University, by Prof. A. Guidobono Cavalchini of the Institute of Agricultural Mechanics of the University of Udine and by Mr. M. Lazzari, a researcher at the same Institute. The study contains a lengthy and comprehensive list of all existing agricultural machinery and current farming practices, and the energy-
saving potential of each.

The study aims to give a preliminary summary of the various types of
energy-saving potential: although some aspects need to be studied in
greater detail for a more complete picture, it is nevertheless a useful
tool for deciding the Communuty's approach and the basic priorities for
action at Community and national level.

Finally, I would like to take this opportunity to thank Dr. P. Abbaticchio
and Dr. R. Gabellieri for their valuable contribution in drafting the
final document and Dr. H.E. Williams for his painstaking work on the
English version of the text.

G.L. FERRERO

C O N T E N T S

INTRODUCTION AND SUMMARY

The aim of this work is to check the possibility of substantial energy savings in the European agricultural mechanisation. In this analytical survey the possibilities of energy saving in stationary plants nor the indirect savings in chemical inputs are considered.

The analysis has been essentially bibliographical, without any direct experimental analysis.

After some general considerations on the European farming structure and the present energy requirements of the sector, the European agricultural machines and tractors industry is outlined.

At the start of this analytical survey, the working schemes and the energy requirements are examined for the different crops. The evolution of tractor manufacturing and the derived machines is also surveyed, gathering the specific implements into the main groups: tractors (and derived machines); soil tillage machines, intercultivation machines and harvesting machines. The evolution and the energy saving potential in tractors and farming machines management is examined and the actions for development are outlined.

A cost/benefit analysis justifies the interest, for the EEC, to invest in agricultural energy saving and to extend this type of analysis to agricultural machinery not considered in the field operations.

1. - GENERAL CONSIDERATIONS

1.1. - Farming structure and mechanization

EEC agriculture is probably the most intensive in the world. The total cultivated area (arable and permanent crops) is estimated to be 79 million ha (table 1). With an EEC population of 319 million inhabitants, this figure corresponds to an average of 4 inhabitants per cultivated hectare. This value is double the corresponding average for the USA, three times higher than for the URSS, and five times higher than for Argentina. Moreover, the EEC farming is essentially based on small-medium size units (fig.1), only 12-17 ha each, 10 times smaller than in the USA (in 1980, the average EEC farm size was 15.7 ha).

This situation, related to the need to supply food and other raw materials for so many people by means of a structure scattered in different climates and various farming systems, induced the increase of mechanization to high levels, sometimes reaching supersaturation. Therefore the load on farm tractors and machines is very high (table 2) with a strong influence on the production cost as well as on the GSP (Gross Salable Product).

This assessment represents a general approach, with particular situations varying from one country to another.

Table 2 shows that the tractor density (tractors/100 ha) is only 3 for Portugal but 21 for The Netherlands. Combine harvesters also show this type of diversity: densities range from 0.4 combines/100 ha in Portugal to 2.8 units/100 ha in The Netherlands. Moreover, even inside each country some differentiations are observed. Fig.2 gives the combine density (units/100ha) for some homogeneous cereal farming areas in France, Italy, Spain, Greece and Portugal. Consider, for example, France, the combines density varies from 0.3 (Landes) to 2.6 (Aisne).

Finally recent calculations show that the mechanization costs, in Italy, in 1985, represent 15 to 16% of the GSP, while the energy consumption represents 5%. Considering the distribution of tractors and combines given in table 2, the figures for Italy could be assumed to be an average roughly valid for the whole European Community.

table 1

Population and harvested area in the EEC Countries

Country	Population	Arable & permanent crops area	Population density	Farms			
				<10 ha		>10 ha	
	inh x 1000	ha x 1000	inh/ha	n°x1000	%	n°x1000	%
Belg.& Lux.	9,855	847	11.6	29	29	70	71
Denmark	5,118	2,830	1.8	27	22	97	78
France	54,219	18,950	2.9	402	35	747	65
(W)Germany	61,638	7,517	8.2	454	53	401	47
Greece	9,793	3,920	2.5	669	91	62	9
Ireland	3,483	977	3.6	71	31	154	69
Italy	56,639	12,450	4.5	1,878	85	314	15
Netherlands	14,310	862	16.6	62	45	75	55
Portugal	10,056	2,779	3.6	695	90	80	10
Spain	37,935	16,521	2.3	1,950	76	621	24
U.K.	55,782	6,909	8.1	70	27	191	73
Totals & averages	319,185	78,569	4.0	4,357	65	2,202	35

(Source: EUROSTAT & direct statistic documentation)

table 2

Number of tractors and combines in service and its evolution
in the EEC

Country	Tractors				Combine harvesters			
	1974	1984	growth	den-sity	1974	1984	growth	den-sity
	n°x1000		%	n° 100ha	n°x1000		%	n° 100ha *
Belg.& Lux.	105.7	115.0	9	14	10.2	9.6	-6	2.4
Denmark	185.5	170.8	-0.8	6	43.2	35.5	-18	2.0
France	1357.5	1535.0	13	8	154.0	148.0	-4	1.5
(W)Germany	1450.5	1471.7	3	20	175.0	165.0	-6	3.3
Greece	93.3	165.0	77	4	4.9	6.5	30	0.4
Ireland	113.0	148.0	31	17	5.0	4.5	-10	1.1
Italy	821.9	1169.5	42	9	27.8	39.1	39	0.8
Netherlands	156.7	188.0	20	21	6.8	5.7	-16	2.8
Portugal	45.4	82.7	82	3	3.9	4.4	13	0.4
Spain	378.5	592.4	56	4	39.7	44.0	10	0.6
U.K.	484.5	529.4	22	7	60.8	56.8	-7	1.4
Total & averages	5172.7	6166.7	16	8	531.4	519.1	-2	1.4

* calculated on cereal harversted areas only
(Source: EUROSTAT & direct statistic documentation)

Fig.1 - Average size of EEC farms (1975)

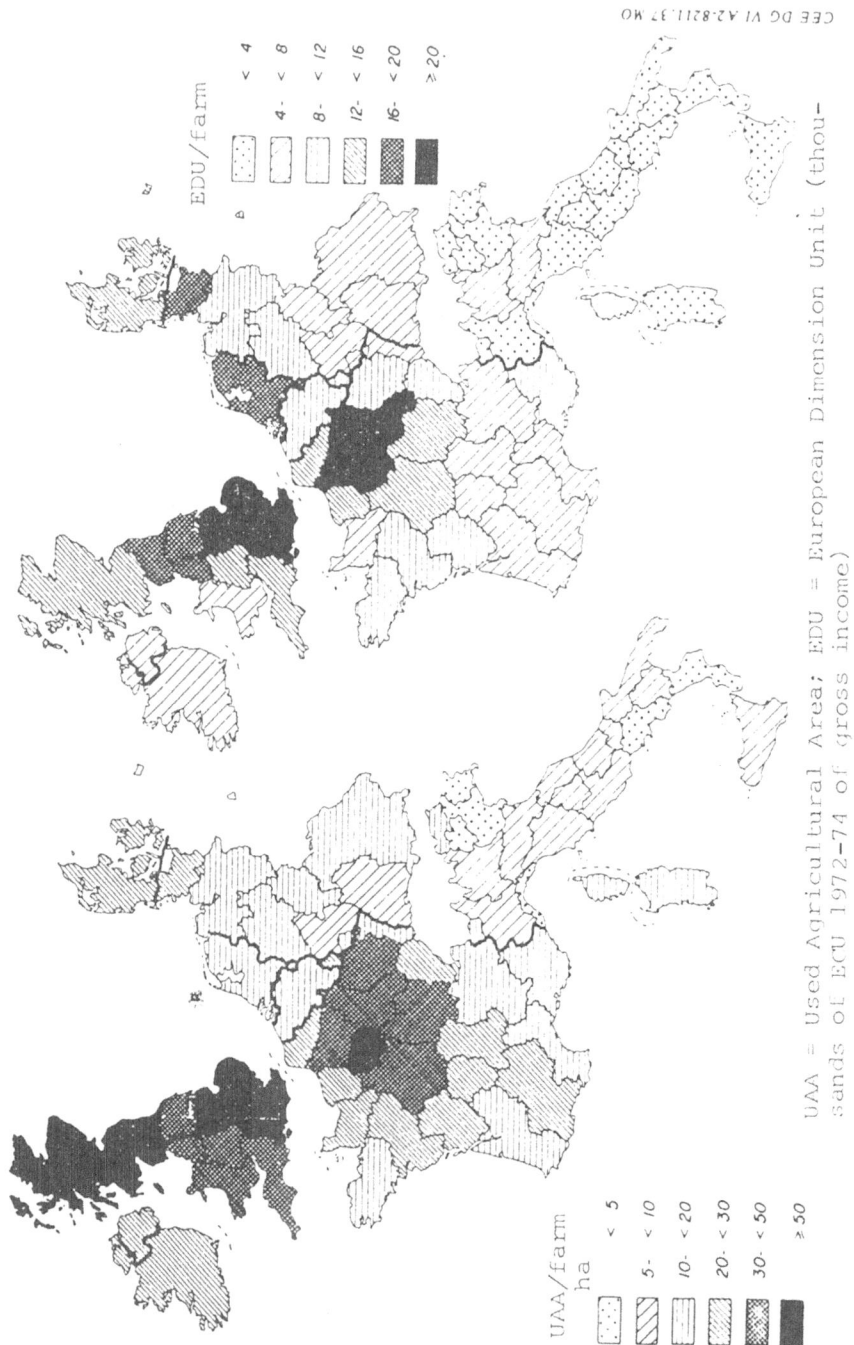

Fig.2

Combine harvester density in 5 EEC Countries

The regions are identified by the province initials
or by the department numbers as in automobile number-plates

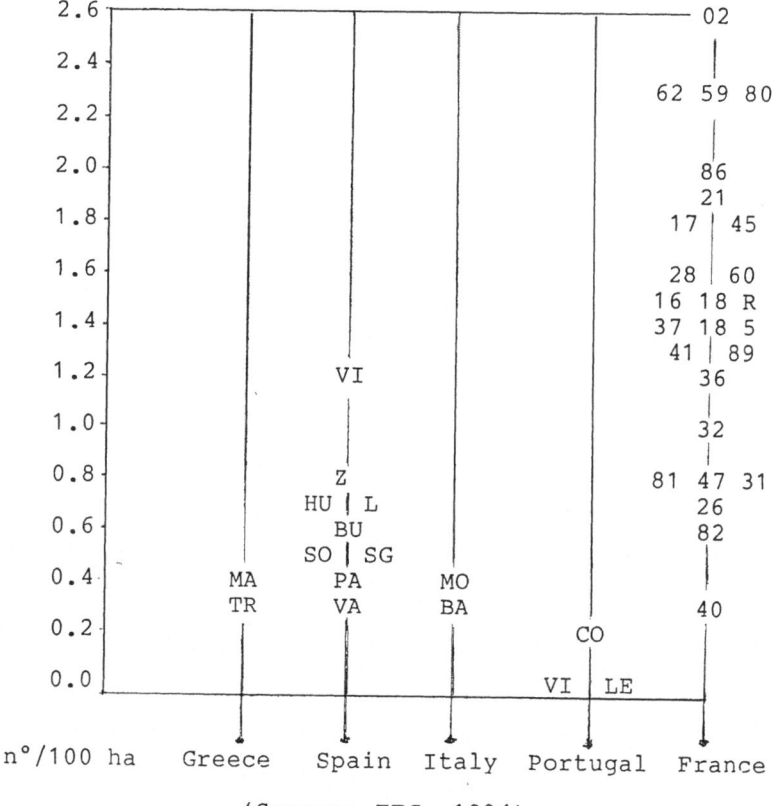

n°/100 ha   Greece   Spain   Italy   Portugal   France

(Source: EEC, 1984)

1.2 - Energy consumption

Table 3 gives the agricultural energy consumption (expressed in tons of oil equivalent) in the European Countries. Basic data are referred to 1977 and, when possible, we have adapted them to 1985.

table 3

Agricultural energy consumption in EEC

Country	Consumptions 1977 (EUROSTAT)		up-to-date (1985) figures	
	ktoe/y	kgoe/ha (*)	ktoe/y	kgoe/ha (*)
Belgium & Lux.	355	430	–	–
Denmark	1,064	401	916	389
France	2,868	153	–	–
(W)Germany	1,250	167	1,260	167
Greece	680	173	–	–
Ireland	185	190	–	–
Italy	1,945	159	2,500	213
Netherlands	603	350	–	–
Portugal	50	14	246	88
Spain	600	29	1,700	83
U.K.	1,383	198	984	132
Total & averages	10,983	139	–	–

(Sources: national statistics and documentation)

The lowest figure is 14 kgoe/ha (*) (Portugal); the highest is 430 kgoe/ha (Belgium). This last value is considered to be exceptionally high and therefore, it may include the consumption for stationary plants.

Following recent EEC data, the average agriculture energy consumption (direct and indirect) totals up to 590 kgoe/ha. Approximately 55% of this total must be considered as indirect consumptions (fertilizers, pesticides and herbicides, animal feed, machinery construction, etc.), while the remaining 277 kgoe/ha represents the direct electric and fuel energy consumption.

The specific consumption (per hectare) given in table 3 for the EEC is higher than the consumption recorded in other industrialized countries, like the USA, where the average consumption (per ha of arable land) is around 180 ktoe/year. However, some of the figures in table 3 appear to have been underestimated.

(*) kgoe = kg of oil equivalent; 1 kgoe = 10,000 Kcal = 41,900 kJ

In addition, the 1977 EEC data for the 9 countries of the Community, indicate the total energy consumption to the farm gate (often including household needs), to be 17.86 Mtoe, corresponding to 348 kgoe/ha.

This is due to the fact that often only the tractor (and other moving machines) fuel is taken into account, without considering the electric energy consumption and the utilisation of energy from raw materials of a renewable nature (agricultural wastes). The last two energy sources are the most difficult to evaluate, because of the scattered distribution of farming activity and the system of "mixed contracts" for the electric energy supply (i.e., contracts in which the professional and the household uses of electricity are combined). As an exemple, in Italy, only 387,000 contracts are recorded by ENEL in farming activities as "professional" (with a total consumption of 1,712 GWh/y), whereas more than 3,000,000 farms are connected with the national grid with a "mixed contract". The "professional" utilization of those mixed contracts has been estimated to be almost 2/3 of the total, so that the total consumption of the agricultural sector can be estimated to be more than 6,000 GWh/y.

Moreover, in Greece, the electricity consumption of the farming activities is officially assessed, in 1983, to be 0.6 GWh/y: this figure appears to be an underestimate.

Finally, in Spain, statistical data indicate an electrical energy consumption of only 1.1 GWh/y.

Even if these figures indicate the order of magnitude of the EEC energy consumption in farming activities, at present it is impossible to establish a detailed picture of the situation. Consequently it would be useful and interesting to fix a common procedure to be utilized in all the 12 countries in order to get a better analysis and knowledge of the situation.

Of the direct consumptions mentioned above, 55 to 75% is due to the mechanical field operations carried out by tractors, tractor derived and other self-propelled machines. The remaining fraction is accounted for by stationary plants and equipments. It is not within the scope of this survey to take into account stationary plants energy consumption in farming activities, even though this is an important sector, both in terms of the magnitude of consumption and the potential for energy savings.

In order to have a reference figure, the direct field energy consumption from the existing stock of working machines has been evaluated.

In the 12 EEC Countries 6,000,000 tractors are operating, having an average nominal power of 40 kw. If we consider the average utilization of a tractor to be 300 hours/year with a load corresponding to 50% of the nominal power (equivalent to a specific consumption of approx. 260 g/kWh), we obtain a total consumption of about 9.4 Mtoe/y.

To this figure must be added the consumptions due to self-propelled equipment and powertillers as well as for irrigation. For the first type of machines, we consider a stock of 540,000 combine harvesters with a total consumption of 0.5 to 0.6 Mtoe/y. For all the other self-propelled machines (including powertillers) statistical data are not available, but we believe - on the basis of the consumptions recorded for some countries - that a consumption of 0.5 Mtoe/y can be considered. The irrigated area in the EEC is around 10,000,000 ha, and considering a specific energy consumption for irrigation of 70 kgoe/ha.y, we get a total consumption of 0.7 Mtoe/y.

The total direct field energy consumption is about 11-12 Mtoe/y. Thus, the average specific energy consumption is approximately 140-150 kgoe/ha for cultivated land. This value appears to be both important and reasonable.

Taking into account all the above mentioned considerations we can keep in mind that the average direct field energy consumption (fuels), in the EEC is approximately 150 kgoe/ha.y. As this is an average value, large variations are possible (sometimes in a range 1 to 6), between different regions, depending on the degree of cultivation intensity, the size of the structures, the existing farming systems, and the socio-economic development levels.

1.3 - The agricultural machines and tractor industry

Concurrently with the widespread farming mechanization, a sound and well developed structure for the production of agricultural machines and tractors exists in the EEC countries.

Table 4 has been prepared on the basis of data directly received from the national associations of manufacturers and from Eurostat. It appears that about 1,800 manufacturers presently exist in Europe, with 150,000 employees. To this some thousand of handycraft units have to be added even though their contribution to total production is small.

table 4

Production of farming machines in EEC

Country	Total production		Tractors		Harve- sters		Tractor compon.		Manu- fact.	Emplo- yees
	mill.tons	ECU x1000	mill.tons	ECU x1000	mill.tons	ECU x1000	mill.tons	ECU x1000	n°	n° x 1000
Belgium	493	140	–	–	493	140	–	–	28*	6.4*
Denmark	209*	60	–	–	209*	60	–	–	47*	5.0*
France	2087	596	591	125	1273	421	223	50	475	30.5
Germany	3349	956	1398	297	1657	597	294	62	189*	47.6*
Greece	28*	8	–	–	28*	8	–	–	36*	1.2*
Ireland	–	–	–	–	–	–	–	–	–	–
Italy	2691	769	1101	234	1153	443	437	92	212*	30.2*
Netherl.	306	87	–	–	306	87	–	–	94	4.9
Portugal	–	–	–	–	–	7	–	–	74x	–
Spain	324	92	205	44	126	48	–	–	–	–
U.K.	1987	567	1006	214	403	232	576	121	667§	23.6§
Total	11471	3275	4301	914	5612	2036	1530	325	1822	149.4

Source: CEMA '84 and national statistics

§ = national data; * = EUROSTAT (more than 25 empl.);x = total

Total production of agricultural machinery is approx. 3.3 millions tons/year, having a value of about 11.5 billions ECU. This figure increases up to 15 billions ECU if we include the handycraft units production.

The tractor manufacturing sector is concentrated in 5 countries and represents 40% of the sales and 30% of the total production.

Because reliable statistical data are not readily available, table 5 only shows the EUROSTAT data for companies with more than 20 employees.

table 5

Production of tractors and agricultural machines in EEC
(1981: manufacturers with more than 20 employees)

Country	Total production		Manufact.
	millions ECU	tons x 1000	n°
Belg. & Lux.	395	112	28
Denmark	209	60	55
France	1,964	561	277
Germany	2,400	685	191
Greece	28	8	36
Ireland	–	–	
Italy	2,392	682	227
Netherlands	1,764	504	418
Portugal	–	–	–
Spain	–	43*	–
U.K.	1,644	461	107
TOTAL	10,796	3,116	1,339

* tractors only

(Source: EUROSTAT)

It is difficult to evaluate the situation for each country taking into account also the small handycraft units as well as the concentrations. Nevertheless, the production given in table 4 is likely to represent around 80% of the total, the remaining 20% being ascribed to small handycraft units, that are very active in some countries in the construction of tractor implements and stationary equipments, but escape the statistic analysis.

As an example of this situation, we consider it would be useful to give some details on the Italian situation.

The Italian annual tractor production, from 1950 to 1983, has increased about 11 times if we consider the number of machines, and 14 times if we refer to the weight. In fact, in the same period of time, the average power of the single units has increased by 70%. Therefore, the tractors production, still limited in 1950 to 20,000 t/y, reached 280,000 t/y by 1983.

In 1950, the sector of tractor implements and stationary equipment was stronger (already producing animal drawn machines): the production was 55,000 t/y in 1950 and increased to 390,000 t/y by 1983 (consequently at present the sector shows 7 times higher). Meanwhile, the types of machines produced has changed and the production has became more diversified.

Both the sectors from 1980 onwards have undergone a marked recession, because of the general economic crisis, the decreased purchasing power of the farmers, and the strong reduction of governmental incentives for farm mechanization. The latter was operating, in 1960-1970, for 40% of purchased machines and is now reduced to slightly less than 7%. In addition, in recent years a reduction in technologic innovation was recorded, but this trend seems to have declined today.

To the two above mentioned sectors, the production of single components and spare parts must be added: from about 4,000 t/y in 1950, it now reaches 80,000 t/y, corresponding to 28% of the production of tractors.

At the same time the market has become more and more international: the export of tractors has risen from 25% to almost 50% while the export of implements and equipment has risen from 10% to 30%. This positive export balance of 1,280 billion liras represents 31% of the value of the total production (4,500 billion of liras in 1983).

48% of tractor exports and 51% of implement exports are to other EEC countries, which indicates that the agricultural machines producted are mainly orientated toward the so called "continental crops", typical of the temperate areas.

All these goods are produced by about 2,000 manufacturers (1,800 handycrafts) with 85,000 employees.

Though EEC manufacturing activity appears rather concentrated in the sector of tractors, tractor derived machines, engines, and self-propelled equipments; it appears extremely scattered in the field of tractor implements and stationary equipment. As a whole, 80% of the production is concentrated in about 10% of the manufacturers. In fact, more than 1,500 producers have sales less than 2 million ECU; 300 have sales from 2 to 14 million ECU and, among the remaining 200 enterprises, there are about five companies with sales higher than 140 million ECU.

A lot of small sized operators in this field, particularly in the sectors of tractors and self-propelled machines, could be considered more as assemblers of single components supplied from outside than as real manufacturers. Consequently, the added value is small.

Geographically, only 0.2% of tractors and 4.5% of the implements production is located in the South.

In more detail, there are 58 tractors manufacturers, the 10 largest having 94% of the market (the first three have 64%) while 20 companies produce less than 20 units/year. There are 18 companies producing engines, the first three covering 75% of the total production.

In the sector of self-propelled machines (powertillers, combines, mowers etc.), 205 manufacturers are listed and the 10 largest of them cover more than 60% of the market. In practice, however, many companies operate simultaneously in several related sectors (i.e., tractors, powertillers etc) so that in both sectors of engine-powered and of self-propelled machines no more than 200 manufacturers are present. There is a large range of models offered: for example, the first 30 manufacturers of tractors offer 750 models and, on average, 25 models are proposed by each producer, with a maximum up to 60.

As far as implements and stationary equipments are concerned, 340 companies supply 88% of the production, 81% are in northern Italy and 9% (handycraft small sized type) are in the South.

Therefore, in total, about 90% of the production is concentrated in less than 20% of the manufacturing companies.

The annual production statistics for the most important agricultural machines in all twelve EEC countries is rewieved in tables 6 to 9: over the last 10 years this evolution has been characterized by a strong recession.

The tables show that in the period 1976-1984, the production of tractors has suffered a 32% decrease, falling down from the 312,000 units sold in 1976 to the 211,000 of 1984. The forecastings developed by the manufacturers associations confirm this trend to the beginning of the next decade, indicating that the size of the EEC tractor market wiull be 180,000 units/y.

A similar trend appears for the most important implements. Over the same period '76 - '84, a 43% decrease has been recorded for the combine harvesters. For the self-propelled chopped forage harvesters, the decrease has been 33%.

table 6

EEC market of tractors

Country	Units produced yearly				
	'76	'78	'80	'82	'84
Belgium & Lux.	5,450	5,630	4,450	4,150	3,700
Denmark	10,400	8,570	3,760	3,850	5,820
France	74,560	64,080	58,780	56,820	54,200
Germany	64,320	58,790	45,480	41,380	34,770
Greece	9,600	11,830	10,600	10,200	10,150
Ireland	6,000	5,840	3,310	2,200	2,390
Italy	62,680	61,140	65,030	49,570	44,160
Netherlands	10,200	9,660	5,750	6,910	6,830
Portugal	5,970	5,510	6,000	5,370	3,380
Spain	30,100	37,400	33,460	21,650	20,750
U.K.	38,200	32,100	21,240	26,100	25,300
Total	312,000	288,720	257,860	228,200	211,450
Index values	100	92	83	73	68

table 7

EEC market of combine harvesters

Country	Units produced yearly				
	'76	'78	'80	'82	'84
Belgium & Lux.	230	270	210	180	170
Denmark	1,970	1,420	1,150	520	700
France	5,630	5,390	4,660	4,250	3,750
Germany	4,350	5,850	4,700	3,200	2,900
Greece	480	300	430	350	200
Italy	2,240	2,100	2,100	1,650	1,390
Netherlands	80	180	130	80	80
Portugal	320	160	150	160	90
Spain	2,270	1,550	1,500	800	800
U.K & Ireland	3,280	3,000	2,510	2 ,280	2,700
Total	21,678	20,220	17,540	13,470	12,060
Index values	100	93	81	62	56

table 8

EEC market of self-propelled chopped forage harvesters

Country	Units produced yearly					
	'79	'80	'81	'82	'83	'84
Belgium & Lux.	85	65	55	80	80	80
Denmark	40	15	15	2	1	3
France	717	650	670	880	690	554
Germany	320	260	220	250	390	350
Greece	-	-	-	-	-	-
Italy	200	220	170	120	120	110
Netherlands	120	125	100	120	115	110
Portugal	-	-	-	-	-	2
Spain	15	15	15	10	5	5
U.K. & Ireland	170	130	120	40	40	25
Total	1,667	1,480	1,365	1,502	1,441	1,240
Index values	100	89	82	90	86	74

table 9

EEC market of round balers

Country	Units produced yearly					
	'79	'80	'81	'82	'83	'84
Belgium & Lux.	40	45	80	100	120	130
Denmark	100	220	150	180	200	170
France	2,130	3,700	5,200	6,800	9,500	10,630
Germany	610	750	900	1,000	1,540	1,700
Greece	-	-	5	5	5	5
Ireland	40	100	100	120	100	-
Italy	500	850	1,650	2,500	3,000	3,800
Netherlands	2	5	5	20	150	200
Portugal	5	5	5	5	2	5
Spain	35	50	40	40	60	80
U.K.	570	750	800	1,500	1,600	2,000
Total	4,032	6,475	8,935	12,150	16,177	18,720
Index values	100	161	221	301	401	464

The trend for round balers is anomalous, because the number of machines sold in the same period has multiplied by four. Nevertheless this is a special case, because the product is new and just reaching his maturity phase: considering the year 1985 as a top sales record, during the current year a 10% decrease is foreseen and this trend is likely to be confirmed in the future.

In response to this situation, the main manufacturers are having to fight for their market share, compressing the manufacturing costs and sales expenses.

The development of full-lines makes some savings in manufacturing costs possible. This production process, adopted since 1960 by the U.S. industry, is now getting a foothold also in Europe as shown, for example, by Fiat-Agri and Deutz.

At first, this tendency may appear unsuitable for the very differentiated requirements of the farming system whose specificity is inclined to demand a special machine for each particular situation and for each crop, just the contrary of the fully standardized machines coming out from a full-line.

Nevertheless, the modern process techniques (CAM = Computer Aided Manufacturing) allow companies of sufficient size to be able to afford them, to get products both differentiated and mass-produced. The development of full-lines makes some energy savings possible via a more rational coupling between implements and tractors.

The decrease in demand for agricultural machines has produced a decrease in R&D investments, therefore limiting product innovation. R&D is now limited to innovations aimed at improving performances, often derived from other industrial sectors, and it is not likely to assist, even in the next coming years, to the introduction of basic innovations. This fact represents an advantage for those manufacturers who are able to update their production. Companies with obsolete equipment are likely to face increasing difficulties, and some will probably disappear or undergo a major reorganisation.

A thrust towards technologic innovations in the agricultural machines industry could be activated only by radical changes in the agricultural structures, and this is not possible in the short or even medium term. New concepts and new processes, linked with the direct (fuel) or indirect (fertilizers, pesticides, seeds, water, etc.) energy savings could also bring a stimulus for innovation.

1.4 - Potential energy saving in field mechanisation

The energy consumptions summarized in § 1.2 correspond to the energy supplied to the powering machines and then transferred to the implements in order to . overcome either the resistances inherent in the operations and in the motion of the machines, or the external resistances the machines meet when operating (soil, crops etc.).

Nevertheless, a proportion of the total energy consumption in the farming activity is also due to some non mechanical factors, related to the users and originated by:

- inadequate maintenance, with a resulting decreased efficiency of the machines;
- tendency to use oversized machines, mainly for tractors in respect of the real needs of the implements; consequently the engines run at a fraction of their nominal power and the specific consumptions are higher;
- the machines are not always skillfully driven, so that an increase in consumption results;
- sometimes the structures (shape and size of land plots, scattered location of the fields in respect of the farming center, road conditions and layout etc.) are not suitable for the mechanization, so that time losses increase versus the total utilized time, and unproductive consumptions increase.

It is clear, therefore, that in parallel with action undertaken to reduce the energy consumption at the manufacturing level, action is also necessary at farm level to help the farmers in their choice and management of machines. Both actions are important to reduce consumption: the second offers, in some cases, more opportunities of energy saving than improvements in engineering and planning.

Our task will be limited to field mechanization, without considering the stationary equipment operating in the farming process, even though these type of machines account for about 40% of the total EEC agricultural mechanisation consumption. We will consider only the 11-12 Mtoe/year of direct field consumptions pointed out at the end of § 1.2. Table 10 gives a general picture of the most important crops in the EEC: they represent more than 85% of the cultivated land and it will be for each of these that our analysis will attempt to locate the energy consuming operations and the possible energy savings.

2 - OPERATIONAL SCHEMES AND ENERGY REQUIREMENTS FOR THE DIFFERENT CROPS

Table 10 gives the data for the EEC areas with the more important crops.

table 10

EEC areas with herbaceous crops and arboreous plants
(ha x 1000)

Country	Cereals	Forages	Pota-toes	Sugar beet	Oil bear. plants	Vine	Olive	Fruit	Total
Bel.& L.	393	148	45	117	5	–	–	11	719
Denmark	1677	401	31	74	167	–	–	7	2357
France	9665	5223	205	527	942	1097	40	206	17905
Germany	4993	1104	244	406	232	100	–	49	7128
Greece	1532	361	60	30	179	180	597	142	3081
Ireland	386	588	35	37	2	–	–	1	1049
Italy	5097	2685	140	211	95	1205	1050	969	11179
Netherl.	197	198	160	129	19	–	–	–	703
Portugal	1201	400	134	3	507	210	335	38	2828
Spain	7455	1284	384	209	833	1717	2086	862	14830
U.K.	4038	1851	198	200	222	–	–	41	6550
Total	36626	14213	1636	1943	3203	4509	4108	2049	68287

(Sources:EUROSTAT and nat.statistics)

These crops represent more than 85% of the cultivated land and are:

-cereals, covering 36.6 million ha (46.5% of the total): 45% durum and soft wheat, 34% barley, 11% maize, 2% other less important cereals, 8% rice;
-forage rotating crops (meadows and fodder grass), covering 14.2 million ha (18%);
-tubers and roots (mainly potatoes), covering 1.6 million ha (2%);
-sugar beet, 1.9 million ha, (2.5%);
-oil bearing plants (sunflower, rapeseed etc.), 3.2 million ha, (4%);
-vineyards, 4.5 million ha, (5.7%);
-fruit tree and citrus tree crops, 2 million ha, (2.5%);
-olive-trees, 4.1 million ha (5.1%).

In the following analysis of the energy requirements for the different crops, the farm-gate consumptions will be given in GJ of primary energy. The total consumptions, at the EEC level, will be given in Mtoe.

2.1 - Cereals

2.1.1 - Farming techniques

In the cereals family, it is useful to distinguish three large groups (wheat-like cereals, maize and rice) whose cultivation practices are very different.

The wheat and wheat-like cereals may be sowed in autumn or in spring, with seed drills, able to realise high seeding densities (500-700 plants/m2). This type of crop does not require deep tilling, because their rootage is rather superficial. Nevertheless, in the southern countries and in clay soils, the habit is to till with deep ploughing (30 to 50 cm) in order to have a sufficient water storage during the last phases of the ripening cycle. Weeding is the sole treatment usually practised. Fertilisers are distributed during sowing and later, once or twice, in surface dressing. Usually the crops are harvested by means of combines but in the southern countries, in mountain or hilly farms, reaper-binders are still used. Harvest time is from the end of May up to the end of June in the South, but in northern Europe the crops are harvested at the end of summer. The straw remains on the fields and is either burned, baled or chopped.

Maize has a spring-summer cycle and requires precision sowing, at low density (5 to 8 plants/m2). Tillage is heavier than for wheat-like cereals, and where cattle breeding is practised in the same region, large quantities of manure are utilized. The harvesting of the grains is carried out from the end of August up to the end of the autumn by means of specially equipped combines. The grains are dried in special dryers or, in the less developed farms, in large cases. The stalks are often chopped to facilitate their covering with soil and also to prevent and fight pyralis.

The rice, in Europe, is cultivated in a spring-summer cycle. This crop requires expensive soil levelling so that fields can be flooded. The sowing is broadcast and the fertiliser and herbicide distribution is similar to that for wheat-like cereals. Harvesting, in autumn, is carried out with combines equipped with special tracks. A final drying stage is always necessary.

2.1.2 - Energy consumptions

In the following tables we have gathered the global energy consumptions(direct and indirect) given by different authors.

Wheat-like cereals global energy consumptions

	GJ/t	GJ/ha
USA(Slesser)	2.9-5.4	-
Europe(Slesser)	0.2-1.8	7.2-54
France(Hutter)	barley & soft wheat 3.4-4.3	18-23
	durum wheat 10.8-12.6	20.5
U.K.(Leach)	3.6-4.3	13.7-18.7
Italy(Costantini)	4.7-4.9	18.7-24.0

Among the different operations soil tillage accounts for
55-65% of the total consumption, while harvesting takes
about 25%. For the EEC the range of field energy consumption
for wheat-like cereals varies from 2.5 GJ/ha (durum wheat)
and 4.3 GJ/ha (soft wheat).

Maize global energy consumptions

	GJ/t	GJ/ha
USA(Slesser) USA(Leach)	3.2-12 5.8-5.9	- -
Europe(Slesser)	0.8-10.8	-
France(Hutter)		23.4-40
Italy(Costantini)		25.2-40

For maize, about one half of the global energy
consumption is fuel which represents 12.6 to 16.2 GJ/ha,
including drying. The drying operation alone requires 50 to
60% of the fuel consumption. Tillage utilises 30-40% of the
remaining fuel consumption, irrigation and other cultivation
practices, another 30-40%, and harvesting 18-20%. As a final
figure, the field operations in which we are interested in
(including the electric power for irrigation) demand 4.7 to
5.0 GJ/ha. In arid areas (South Europe) the direct energy
consumption has been evaluated at less than 3.6 GJ/ha.

Rice global energy consumptions

	GJ/t	GJ/ha
USA(Slesser)	6.8-11.9	-
USA(Leach)	10.8-12.6	-
Europe(Slesser)	0.2-10.8	-
Italy(Cavalchini)	5.5-6.5	27-36

The figures for energy consumption in Italy are the most reliable for all the EEC. The direct consumptions included in that value(fuel, lubricants, electric power) are between 11.9 and 12.6 GJ/ha. The drying of rice represents 40 to 50% of this figure, tillage 25-32%, cultivation practices 10 to 20%, harvesting 10 to 15%. For the European culture of rice, the energy consumptions for field operations can be finally evaluated in the range of 5.4 - 7.2 GJ/ha.

Fig 3 summarizes the distribution of energy consumption for cereals. For the whole family, averaging the different species and considering only the energy requirements for the field operations(that is excluding drying and storage), we obtain a direct consumption of 3.6 to 4.1 GJ/ha. Soil tillage represents 55 to 60% of this figure, harvesting operations 25%. For the whole EEC 3.3 to 3.5 Mtoe/year is required to cultivate 36.3 million ha of cereals.

Fig. 3

Cereals, forages, and oil seed crops direct energy requirements

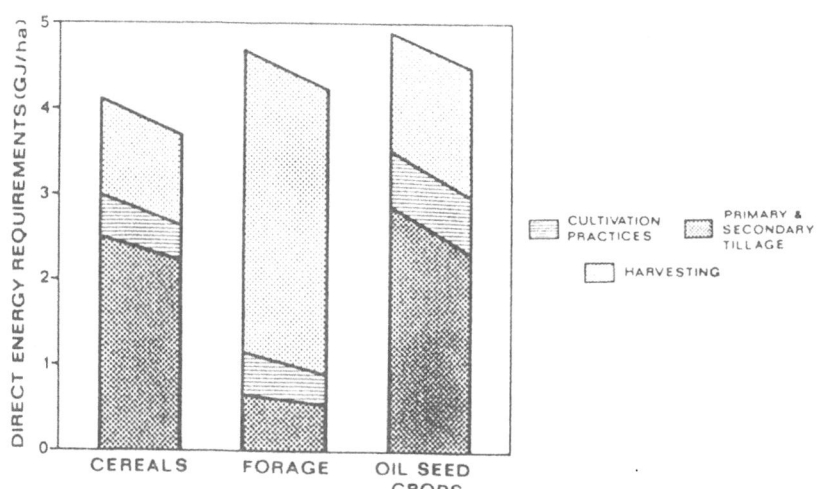

2.2 - Forages

2.2.1 - Farming techniques

The farming techniques for forage crops are very disparate. In this survey only rotating cultures and green maize that is to be ensiled are considered.

The duration of the fodder rotating cultures can vary from 2 years in the short shifts up to 4-5 years. The layout operations also vary with the type of plants: for instance, alfalfa demands deeper tillage than graminaceous crops. The culture \operations carried out in the years after implantation are usually a spring harrow and top dressings.

The harvesting operations vary depending on the working scheme: green forage, traditional haymaking, two-stages haymaking, ensilage, dehydration. The number of annual operations depends on the environment, the plant species and the techniques employed (considering also irrigation). In northern Europe and in the dry southern regions, only two haycuttings are practised for the fodder rotating graminaceae, while up to 5-7 haycuttings are practised for the irrigated leguminous plants in southern Europe.

For green maize the tillage, cultivation practices and the fertiliser distribution are the same as for the maize harvested for grains. If the culture is a "second crop", as is possible in irrigated regions with strong insolation, "minimum tillage" is employed. The green maize is harvested by means of mower-chopper-handling machines and the ensilage is commonly stored in horizontal silos in which the forage is compressed by means of wheeled tractors.

2.2.2 - Energy consumptions

In the following table we have combined the data available from different authors on the energy consumptions of forage crops.

Forage crops energy consumptions

meadow forages	global consumptions GJ/ha	direct consumptions GJ/ha
Leach	9.4-5.2	(low intensity (0.25 (high intensity (2.5-5.8
Hutter alfa-alfa (France) clover sorghum	9.4-33) 28.4-73) 36-54)	4.7-36
Prèche (mowing & harvesting only) (France)		0.7-2.0
Noren with ensilage (Sweden) " roto-baler " drying	3.4 3.8 4.7	- - -
Pellizzi (North. Italy)	3.2-46.8	(irrigated areas (10.8-11.5 (arid areas (5.4-7.2
silo maize		
Pimentel (USA)	50	4.3-8.1
Pellizzi (N.Italy) (Europe)	28.8-39.6 -	3.6 3.6-4.3

Leach's data takes into account the variations due to the production intensities and to the different harvesting techniques (grazing, haymaking, ensilage). For two-stage haymaking, to the 2.5-5.8 GJ/ha for field operations,19.8 GJ/ha must be added for the artificial drying.

In Hutter's data, in the case of alfalfa, the minimum consumption is recorded for dry meadows partially grazed, while the maximum is for haymaking in irrigated areas. The direct consumption represents 30-35% of the total for sorghum silages, 40-45% for clover, 50-70% for grazed alfalfa, 63 -70% for reaped alfalfa.

The Noren's data for Sweden are also valid for Ireland, Scotland and Denmark and are for forages harvested with two cuts per season. In the case of artificial drying, Noren evaluates the drying consumption to be 1.8 GJ/ha.

The figures given by Pellizzi for Northern Italy are for irrigated crops with 4-5 cuts per season. For the fodder maize, the productions per ha considered in Northern Italy by Pellizzi are 2-3 times higher than the unitary productions considered, for the USA, by Pimentel.

Finally, averaging the energy consumptions of all the forage plants grown in Europe, the direct field consumptions for the cultivation and harvesting are approximately 4.3-4.7 GJ/ha: lower figures (2.2-2.5 GJ/ha) are recorded where(Northern Europe and Southern dry regions) only very few cuts are possible. The harvesting operations require, on average, 50-70% of these values: the higher figure is for irrigated areas. Irrigation, if practised, takes 30% of the total. Soil tillage represents only 10-15% because its cost is distributed on several years (Fig.3).

For the whole EEC, the energy field consumption for these crops is approximately 1.2-1.4 Mtoe/year.

2.3 - Sugar beet

2.3.1 - Farming techniques

Generally sugar beet needs a deep primary tillage, though some regional variations do occur, i.e.deeper in clay soils, lighter in the loam soils of central Europe. Great care is needed for the seed-bed preparation, because this influences the ease of emergence of the small plants. Sugar beet benefits from manure, when available. Chemical fertilizers are used before seeding and with row distribution, during the seeding operations.

For precision seeding, precoated single-germ grains are used, in order to avoid a subsequent thinning of the crop. In some regions an earthing up is still practised.

Besides herbicide treatment, a treatment against cercospores may be necessary as well in some irrigated regions.

Harvesting (from mid-August till the late autumn) can be carried out mechanically either in joint or in separated operations: in the last case, two or three field passes are necessary.

2.3.2 - Energy consumptions

The global energy consumption of the beet culture can be summarized in the following table.

Sugar beet global energy consumption

	GJ/t	GJ/ha
USA (Slesser)	1-1.3	-
Europe (Slesser)	0.5-1.2	18-72
Italy (Cavalchini)	0.35-0.45	18.9-21.6

The upper value of Slesser's data for Europe, is only valid when irrigation is practised, and this is not common for northern Europe, where most of the sugar beet culture is located.

The direct field consumptions of fuel and lubricants vary, in the EEC, between 5.0 to 6.1 GJ/ha. Tillage (40%) and harvesting (40-50%) require the highest proportion(fig.4).

Transferring these figures to the EEC level, sugar beet occupies 1.9 million ha of harvested area and the corresponding direct field energy consumption is about 0.25 Mtoe/year.

2.4 - Potatoes

2.4.1 - Farming techniques

Potatoes require the typical operations normally practiced for renewal cultures: ploughing must be sufficiently deep to cover the manure, when available.

Seeding is mostly carried out by means of automatic machines in northern Europe, while in the southern countries semi-automatic machines are used. Mineral fertilizers are supplied before seeding, and in top dressing.

In heavy soils multi-row hoeing is practised but the earthing up is considered the most important operation. Besides weeding, it is also important to combat the potato beetle by means of specific treatments.

In the more advanced areas, the harvesting is carried

out by self-propelled machines or by machines powered by tractors which operate jointly. The first system requires a further stationary sorting point.

2.4.2 - Energy consumptions

The following table gives the data for the global energy consumption of potato culture.

Potato culture global energy consumptions

	GJ/t	GJ/ha
USA (Slesser)	1.6-3	46.8-93.6
Europe (Slesser)	0.2-4	3.6-126
U.K. (Leach)	1.5	36

For the U.K., the direct field consumptions have been evaluated by O' Challagan at 5.8 GJ/ha while, for northern Italy, Cavalchini evaluates the direct field consumptions at 3 GJ/t, corresponding to 7.2-7.9 GJ/ha. In these figures the tillage represents 30-35%, harvesting 50-60% and seeding 8-10% (fig.4).

Transferring these values to the EEC level, the average field energy consumption for potatoes is around 0.30 Mtoe/year.

Fig. 4

Sugar beets and potatoes direct energy requirements

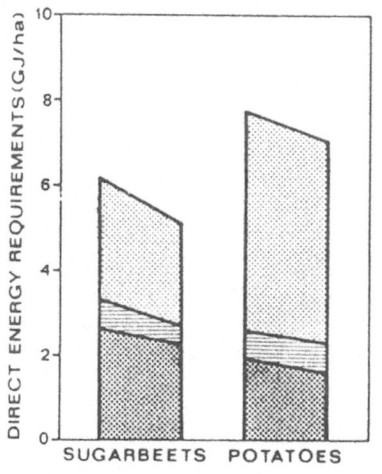

2.5 - Oil seed crops

2.5.1 - Farming techniques

The most important oil seed crops in the EEC are: rape seed (in northern countries); sunflower and soybean (in central and southern regions).

Rape seed in the crop rotation system occupies the same place as wheat: seeding in autumn, after a medium depth ploughing and a convenient seed-bed preparation by means of conventional seed-drills. The fertiliser dustribution is divided into two parts: the first one before seeding, the second as a spring top dressing. Herbicide can be applied either before or after emergence and sometimes a mechanical weeding is necessary. Harvesting is performed by combine harversters.

Soya and sunflower are spring-summer cycle crops with the same requirements as maize as far as tillage, cultivation practices and harvesting are concerned. Harvesting is performed by combines. Both soya and sunflower crops can be cultivated as "second crop".

2.5.2 - Energy consumptions

Oil seed crops energy consumptions

	global consumptions	direct consumptions
	GJ/ha	GJ/ha
Rape seed		
Germany (Thoma)	16.5	3.2-3.4
France (Hutter)	15.5	5
Soybean		
USA (Pimentel)	9.4-15.1	-
France(Hutter) with irrigation	18.7	8.6-9.4
without "	11.5	4.3-5
Sunflower		
France (Hutter)	17.3-28.8	5.4-6.1
USA (Pimentel)	18-54	-
Italy (Cavalchini)	-	4.7-5

In the direct consumption given by Hutter for rape seed in France, 60% is for tillage and 25-30% for harvesting. On an EEC level, the direct field consumption is estimated to be 3.6 GJ/ha, of which 55-65% is for tillage and about 25%

for harvesting.

From the direct consumptions given by Hutter for the soybean culture in France, 2.2 GJ/ha is for tillage, 0.7-0.9 GJ/ha for harvesting. If the cultures are irrigated, an additional 3.6 GJ/ha is required.

To the direct consumptions given by Hutter for the sunflower cultivation in France, 5.8 GJ/ha must be added if irrigation is practised. In Pimentel's data for sunflower cultivation in the USA, the irrigation energy requirement is (when irrigation is used) 32.4 GJ/ha, all the other operations account for 8.8 GJ/ha. Cavalchini's data for Italy are the same as for maize: 50-65% of the requirements are for tillage and 30% for harvesting.

Averaging all these figures for all the oil seed crops and for all the EEC countries, the direct energy requirements range between 4.5-4.9 GJ/ha, 60% is for tillage and 25-30% for harvesting(fig.3).

For the oil seed crops, at the EEC level, the field energy requirements are about 0.35 Mtoe/year.

2.6 - Vineyards

2.6.1 - Farming techniques

European grape-growing is characterized by a great heterogeneity of growing forms and pruning systems: horizontal trellis and small-tree vineyards are spread in warm regions (small-trees are compulsory in windy areas), while hedgerow vineyards, together with other growing forms (wheelspoke, "pergola", double curtain) are typical of the viticulture of temperate-cold regions.

The life span of a vineyard is usually 20-25 years, so that the implantation practices, even if expensive, do not have a great influence on the average yearly energy consumption.

Soil tilling operations are generally carried out mechanically but sometimes a chemical weed-killing agent is preferred or, if possible, (as in France for more than 80% of the vineyard area) a grassy covering is adopted.

Pesticides are applied mechanically by various systems and, on average, 6 to 10 treatments/year are carried out, with maximum applications of up to 20 treatments/year, depending on the environment and metereology.

Generally, pruning and grape-harvesting are still manual operations. Pruning is made easier through the use of small

air-compressors (powered by tractors p.t.o.) to operate
pneumatic pruning shears. Grape-harvesting is aided by the
use of small tractor-drawn lorries to convey the grapes to
the cellar. In France, mechanical grape-harvesting is
already widespread (6,500 grape-harvesters) while in Italy
it is just beginning (300 harvesters).

2.6.2 - Energy consumptions

Baldini evaluates the viticulture global energy
consumption to be between 37-59 GJ/ha. Only 15-25% of this
figure, that is 7.2-10.4 GJ/ha, represents the average
direct energy requirement, but lower direct consumptions are
recorded in the mediterranean plantations, especially for
vineyards of the small-tree form. The direct energy
consumption is divided as follows (fig.5):

- pruning and pesticide distribution   46-65%
- soil tillage   20-40%
- harvesting   5-20%

It must be pointed out that in the case of highly
mechanized plantations, using pruning and grape-harvesting
machines, the field energy requirements are average because
the above mentioned operations represent less then 20% of
the total, as confirmed by the detailed investigations
carried out by Dellenbach and Lacombe in France.

Averaging the above data for the 4.5 million ha of EEC
vineyards, the global direct energy consumption for this
culture is between 1.4-1.6 Mtoe year.

Fig. 5

Vineyards and fruit tree crops direct energy requirements

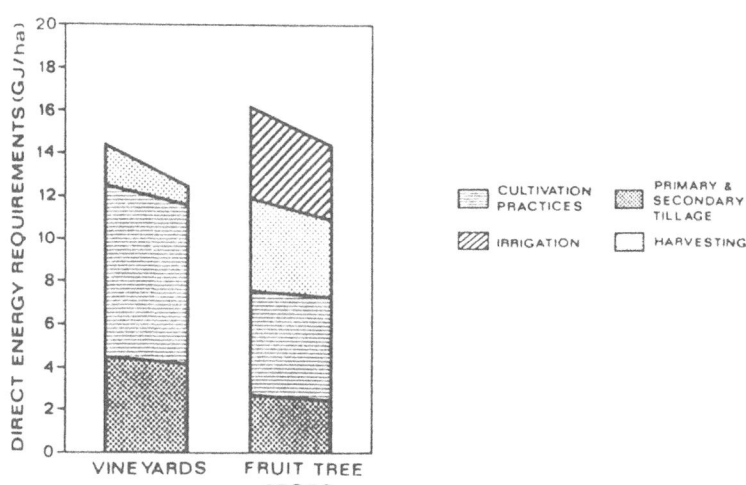

2.7 - Fruit and citrus tree crops

2.7.1 - Farming techniques

A very large number of crops can be classified as orchards and citrus trees. In our survey we will limit the evaluation of the energy consumption to apple, peach and the citrus trees, these being the most widespread fruits in Europe.

The apple-tree is traditionally pot-grown, utilising dwarfing understocks but this technique is being replaced by high intensity plantations with trees grown in rows. The life span of the plantations is generally 20 years, but many examples of trees lasting much longer are recorded. The intense phases of the apple production cycle are pesticide distribution, pruning and harvesting. Pesticide distribution can be repeated up to 20 times per year. Pruning and harvesting are labour intensive. Their mechanisation is limited to the use of pneumatic pruning shears and lorries to collect the fruit. An increasing common operation is thinning, usually carried out manually in spite of the possibility to do it chemically. Other requirements of this crop are irrigation and frost protection operations.

The peach-tree is also traditionally pot-grown or dwarf palm-grown, but the more recent plantations are intensive ones with more than 1000 freely growing trees per hectare. The mechanical soil tillage is often integrated with (or replaced) chemical weed-killing. Fertilizer distribution and irrigation are necessary. Pesticides distribution is practised in winter, spring and summer. Pruning and thinning of the fruit are manual operations made easier by means of auxiliary machines or, for thinning, by means of spraying with drop down agents. The handling of the harvested fruits is increasingly being carried out by means of pallets, particularly when the fruits are for industrial uses.

Nevertheless, for both apple and peach crops, shaking machines are slowly being introduced for mechanical harvesting, which improves productivity and reduces the production costs.

Citrus trees are being developed into more convenient growing forms. Because plantations last 40-50 years, the layout cost is (yearly) very small. The most important factor in the production process is irrigation. The traditional technique (pot-irrigation) is being replaced by sprinkling or drip-irrigation systems: both of these new techniques are more expensive in terms of energy consumption. The annual fertilizer distribution is carried out mechanically. Atomizers for the two pesticide distributions are only used in the most modern plantations.

Soil tillage is carried out by power-tillers or, when possible, by tractors. Pruning is carried out every two years. Harvesting is usually a manual operation, though in the most modern plantations auxiliary machines are used.

2.7.2 - Energy consumptions

The following table summarizes the energy consumptions for apple, peach and citrus trees.

	global consumptions		direct cons.
	GJ/t	GJ/ha	GJ/ha
apple-tree			
USA (Slesser)	-	57.6-104	-
Italy(Baldini et Al.)	-	50.4-64.8	22.3-32.7
peach-tree			
Italy(Baldini et Al.)	-	42.6-57.6	12.6-25.2
citrus-tree			
USA (Slesser)	1.3-3.6	-	-
Italy(Baldini et Al.)	1.6-1.7	43.2-50.4	10.8-16.2

In the apple-tree crops the most expensive operations are harvesting, pesticide distribution and pruning, which represent about 35-40% of the total. Tillage has a low energy demand, mainly because the soil remains grass-covered.

In the direct energy requirements for the peach-tree, harvesting accounts for 30-40%, roughly the same figure as for both pesticides distribution and pruning.

50-60% of the direct energy requirement for citrus tree cultivation is for irrigation, the remaining 40-50% being divided between tillage, treatment and harvesting.

Averaging the data for all the orchards (fig. 5) the energy consumption for orchards and citrus trees is 14.4-16.2 GJ/ha. Harvesting and treatments account for 4.7-5.4 GJ/ha each, irrigation utilises 4.3-4.7 GJ/ha, tillage 2.2-2.5 GJ/ha.

The global consumption for the 2 million ha of fruit + citrus crops in the EEC is estimated to be 0.7-0.8 Mtoe/year.

31

2.8 - Olive-trees

2.8.1 - Farming techniques

Even if some examples of intensive olive-tree cultivation indicate potential high yields, generally traditional methods are used. Therefore the plantations do not take into account the possibilities of modern mechanisation. Moreover, the olive-tree culture is often associated with other crops, like cereals. The time span of the plantations is estimated to be 150 years.

The main crop operations are: tillage (theoretically two ploughings and two harrowings per year), 2-4 treatments, fertilizing (in two steps) and pruning (every two years). Usually the costs of tillage are charged to the main crop (durum wheat) and the cost of tillage for the olive-tree crop is not taken into account.

Harvesting is usually a manual operation, using nets and brushing machines. Shaking machines are used in limited areas of Southern Italy and Spain.

2.8.2 - Energy consumptions

Spugnoli estimates the direct consumption to be 8.3 GJ/ha when harvesting is carried out by means of shaking machines. This figure becomes 3.6-4 GJ/ha if the harvesting is done by traditionnal method. The mechanization of olive harvesting in Europe is not common. In Italy, for example, on more than 1 million hectares, only 400 shaking machines are operated. As a consequence, the average European direct consumption is evaluated at 3.6-4.3 GJ/ha.year.

Fig. 6

Olive tree direct energy requirements

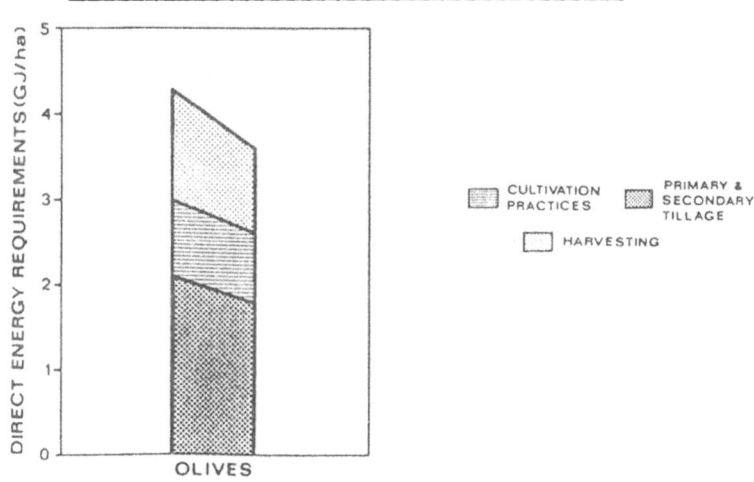

It is likely that, in the future, modifications in plantation techniques and the increase in mechanical harvesting will increase the direct energy consumption up to 9-12.6 GJ/ha.y. Today the average direct consumption is 4 GJ/ha: 50% for tillage and 20% for pruning (fig. 6). The latter operation, even if practised only every two years with the aid of pneumatic pruning shears powered by small power-tillers, remain relatively expensive in terms of energy requirements. Transferring the above mentioned average figure (4 GJ/ha) to the total area harvested in the EEC (4.1 million ha), we obtain an equivalent primary energy consumption of 0.4-0.5 Mtoe/year.

2.9 - Energy consumptions for handling in farming activities

The utilization of tractors and the development of modern farming activities, have been influenced by the increasing importance of transporting products to and from the farm. It has been estimated that in farming activities some 40 t/ha.y must be handled, with peaks up to 120/ha.y and minima at 10-20 t/ha.y.

This is linked with the development of direct contacts with the market and the increase of intensive cultivations. Farm handling can be divided in:
- internal transports,
- transports to and out of the farm,
- external transports.

Part of the energy requirements for internal handling have been detailed in the previous paragraphs. This is not the case for the other types of handling, often carried out by means of mechanical-drawn lorries or other container-carriers. There is a lack of data on this subject. Nevertheless, we have made some basic calculations which must be considered as very rough.

Forages crops are not included in these calculations, because in most of the cases they are used inside the farm. The other crops are mostly (90%) forwarded to storage centers for processing and marketing. On the basis of some Italian experiments, the energy requirement of this type of handling is 4.3-6.5 MJ/t.km. Without considering forage areas, the global handling energy consumption, at the EEC level, for the 54 million hectares harvested, is 0.50-0.60 Mtoe/year.

2.10 - Conclusions

The survey carried out on the literature internationally available confirms the figure (chapter 1) of 11-12 Mtoe/year for the field energy consumption in the EEC agricultural activities.

The crops considered cover 85% of the arable & permanent crop area and we have estimated a total energy requirement of 8.8 Mtoe/year, taking into account that the remaining 15% is covered by vegetables and industrial crops requiring high energy inputs (from 5 to 20 GJ/ha). The corresponding figure for 100% of the cultivated area is slightly more than 11 Mtoe/year.

Fig. 7 gives the histograms for the several cultures and **Fig. 8** shows the sharing of the total energy consumption among the main farming operations.

Fig.7

Total European direct energy requirements for the different considered crops and for transports

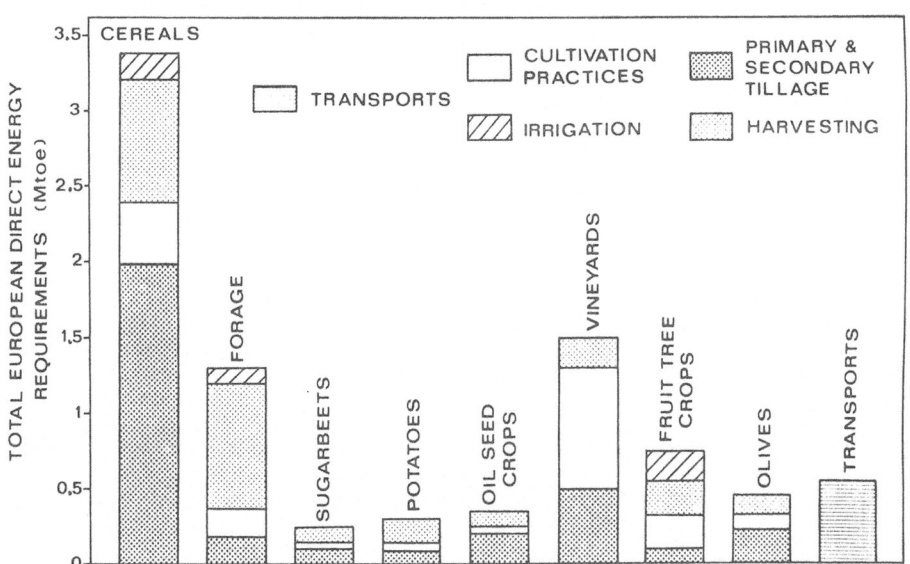

Fig.8

Total European direct energy requirements per operations

11÷12 Mtoe

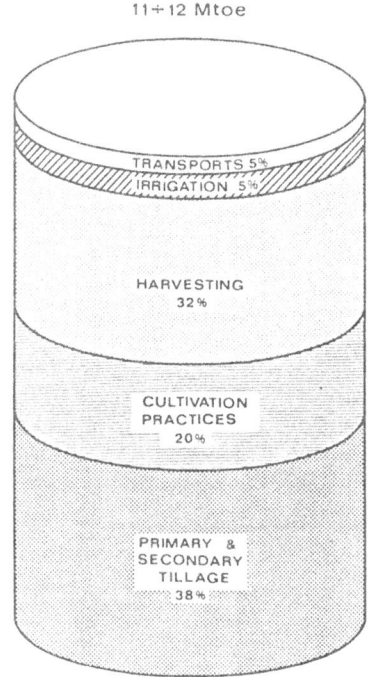

TRANSPORTS 5%

IRRIGATION 5%

HARVESTING
32%

CULTIVATION
PRACTICES
20%

PRIMARY &
SECONDARY
TILLAGE
38%

Taking into account the differences between the various crops, it is interesting to point out the following considerations:
- the soil tillage operations (including - when necessary - the seed-bed preparation), on average, account for 38% of the total consumption;
- fruit tree crops are very demanding in indirect energy requirements while the direct energy consumptions for the cultivation practices represent only 20% of the total;
- harvesting takes 32% of the total energy requirement;
- irrigation represents 5% of the same total;
- transport accounts for the remaining 5%.
    This distribution of energy requirements is important because it shows that, for energy savings, most of the attention should be paid to the primary tillage and harvesting machines. It is necessary to establish whether improvements are possible in the technical and operational efficiency of the machines themselves and also in the work organisation(right choice of mechanizations and updating of the cultivation techniques).

3 - TRACTORS AND AGRICULTURAL MACHINES MANUFACTURING: EVOLUTION AND ENERGY SAVING PERSPECTIVES

Applied research for energy saving in tractors and agricultural machines has always been carefully considered in the planning and construction of new engines and new implements, though in the years 1955-1975 this aspect of the engineering lost some of its interest. After the energy crisis which followed the Kippur war (1973) and the second oil crisis of 1979, both the private and public research have been driven again to search for fuel saving techniques.

The main objective of this activity has been to improve the performances of the different machines; some innovations have lead specifically to energy savings.

3.1 - Tractors and other agricultural vehicles

At the international seminar on the rationalization of farm mechanization held in Verona (March 1977) some lecturers (Manby,UK; Gego,Germany; Millar,USA; Ponzio,Italy) pointed out the numerous possibilities for energy saving in the design phase of tractors, tractor-derived machines and other self-propelled machines. The efficiency of tractors was analyzed in respect to engines, the power transmission and the wheels-soil system. The tactor equipment with power take-off, the coupling with mounted tools and the control of the hydraulic lifting system (in order to avoid or reduce slipping losses) were also discussed.

Table 11 summarizes the points, in the planning phase, that may reduce the energy consumption during the use of the machines.

-Engines: even though modern Diesel engines have achieved good efficiency, the target is to reach a 40% efficiency. This can be done by acting on the feeding system (supercharging for high-power engines), on the lubrication system and on the cooling devices. Higher accuracy in construction, a standardization of the components and a reduction of their number, are considered to be the basic elements for efficiency improvements in existing engines, without significantly increasing the farming costs.

table 11.

Values obtained in practice and prospects of improving efficiencies of tractors

Components	Range of action related to max. output %	Values obtained in practice (dimensionless)	Possibilities of efficiency improvement (loss reduct.)
Engine	0-100	0.28-0.37	Ltd. possibilities by: exhaust turbine injection feed. speed, etc.
Mechanical change-sp. gearbox	0-100	0.85-0.95	negligible
Driving power transmission	0-100	0.40-075	by all-wheels drive, radial tyres,favourable operating conditions, shifting to PTO shaft power,differential lock,axle load distribution, etc.
PTO shaft	0-100	0.9-0.97	Negligible
Hydraulic (power lift, steering,etc.)	5-35	0.55-0.75	Ltd.possibilities
Residual power (heating, pneumatics, etc.)	0-15	0.5-0.75	Ltd.possibilities

(Source:Gego,1977)

Transmissions: present efficiencies are very high (0.85 up to 0.95), so that little margin remains for further improvements. However, the use of epicycloidal gears in synchronised gear boxes allows a couple-shift under load; a higher precision in the construction of gears and the coupling of the mechanical gear-box with a hydrodynamic transmission; the use of hydraulic controls; are all factors contributing to the improvement in operational efficiency of the tractors, the comfort of the driver and, finally, the quality of the work. It has also been recommended that devices are installed to monitor the load of the tractor: the driver could then adapt to the load the engine charge.

It appears that the use of hydrostatic trasmissions, even if coupled with mechanical gear-boxes, are now obsolete. The use of this type of device will be limited in the future to very specific equipment, where the resulting advantages can justify the above mentioned limitations.

Wheels-soil systems possess a significant potential in terms of improvements in tractor efficiency. Presently the values recorded for the field efficiencies vary from 0.40 to 0.60 and considerable improvements are possible:
-through the type of the tyre (for ex., using radial-ply tyres);
-developing models with 4 wheel-drive, which utilise the weight of the tractor itself, without the need for ballasts;
-improving the hydraulic system for the automatic load control.

It is considered possible, through these developments to reach field efficiencies of 0.70-0.75.

Another possibility, not yet thoroughly investigated, is the use of wider tyres at lower pressure.

Besides design improvements and construction, other factors appear important if energy savings are to be fully realised: good maintenance of the tractors, proper organization of their use, compatibility between tractors and equipment, more suitable and widespread utilization of the power take-off and of the front- or rear- mounted implements.

These were the conclusions of a meeting held 8 years ago. We will indicate which achievements have been reached and those that as yet remain to be attained. In view of the possible energy savings, we will consider four significant subjects: engines, tyres, electronic equipments and tractor architecture. A brief discussion will be made about the replacement of the diesel-oil with alternative fuels.

3.1.1 - Engines

In Europe, diesel engines have been used for many years on tractors, on other agricultural vehicles and also on self-propelled machines for field operations.

The present diesel engines have efficiencies of 35-38%, to compare with the typical values of 32-36% recorded in 1977. Therefore fuel saving has been obtained. This performance is due mainly to the increase in the compression ratio and of the specific volumic power (kW/dm3 of cubic capacity) and also to the improvement of the injection systems by using rotary pumps and high atomization injectors. Engines of more than 50 kW are usually supercharged: all the European manufacturers offer this type of engine which are 10-15% more efficient.

Many other energy saving factors remain as targets for the coming years. Research in this sector is today (from the standpoint of the energy saving) examining a combination of factors: useful lifetime, low consumption, good adaptation at variable loads, easy starting at low temperatures, low maintenance. Within this framework there are many possible solutions for greater efficiency. One of the most important solutions is the improvement of fuel intake into the combustion chamber, so that a more complete combustion takes place, resulting in higher power. The result of this line is to carry out research towards turbocharging of small engines.

Fig. 9 gives the scheme of a turbo-charging system with intercooler.

Fig.9

Turbo-charger with intercooler system

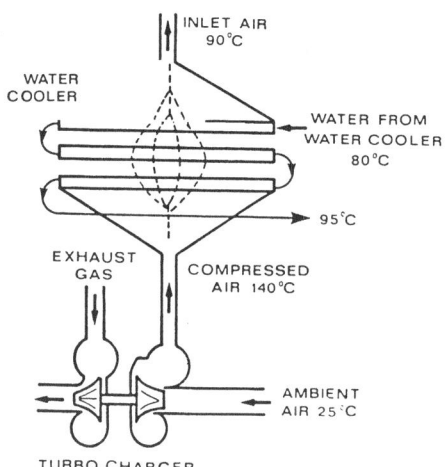

This system recovers a proportion of the thermal and kinetic energy of the exhaust gas and uses it to compress the air entering the combustion chamber thus increasing its density. Through this system, the power of the engine can be increased by 20-25%, with a parallel decrease in the specific fuel consumption. In the latest models, consumptions as low as 193-196 gr/kWh have been obtained. This figure represents a reduction of 20-25 points versus conventional models and efficiencies up to 44% have been attained (Fig.10).

Fig. 10

History of fuel consumption rate
(Source: Kihara, 1985)

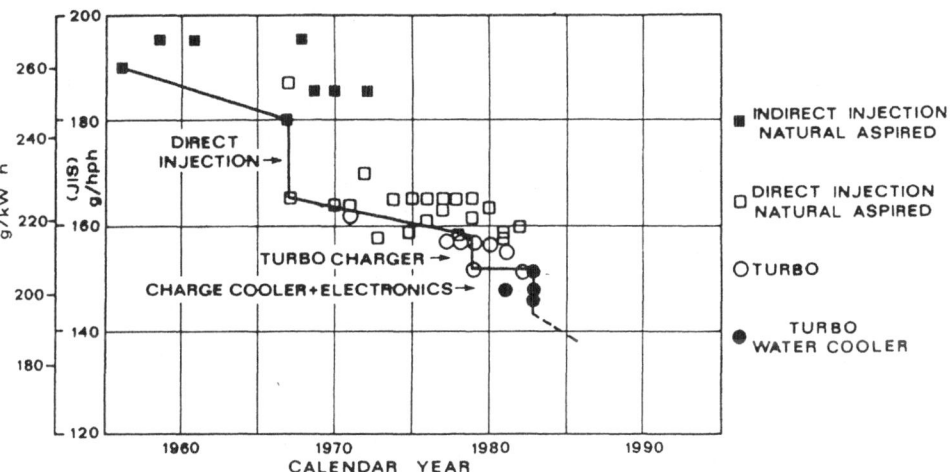

This trend in the reduction of fuel consumption is likely to continue in the coming years: Jenny reports that, in diesel engines having a low rotating speed, a 50% efficiency has been exceeded.

Probably, following the provisions of Clowings (Perkins Co.), this development will be carried out in a different way for three power classes: low (less than 50 kW), medium (50 to 75 kW) and high (more than 75 kW) power engines. This distinction is not likely to interfere in the export of engines to developing countries, where the demand is for cheap equipment that is easy to operate and maintain.

The 50 to 75 kW class represents the most important market segment for the European manufacturers, which is orientated towards four cylinder, four litre engines, naturally aspired or turbo-charged. The turbo-charged types are likely to become more widespread, as the market trends indicate the replacement of 6 naturally aspired cylinders by 4 turbo-charged cylinders. The improvement of these engines is linked to the availability, during the planning phase, of sophisticated tools necessary to develop the engineering solutions (shape of the combustion chamber, injectors location, etc.). The availability of digital computerized calculation systems (finite elements, computer assisted drawing etc.) will facilitate the evaluation of these potential solutions.

The third class of engines (more than 75 kW) is for tractors typically used on large farms or by utility companies. Therefore, they must have high performance in terms of efficiency and ergonomy, as they are typically operated for 2000 hours/year. In other words, the quality of these engines must be higher in comparison to the engines of the 50 to 70 kW class described before. On the other hand it is relatively easier to equip this type of high powered (and expensive) machine with sophisticated control devices, the cost of the auxiliary equipments being easily absorbed.

Among the possible innovations, the use of <u>electronic control systems</u> acting directly on the injection pump (Fig.11) seems the most likely in future years. This system is already being examined by the larger manufacturers. Its function is to connect the load of the engine with the pressure and the flow of fuel in such a way as to optimize the engine performance. This system will be one element of a more complex control system linking the most important operational parameters, and actuating strategic programs for the optimization of the global and/or specific efficiencies.

Besides the above mentioned points, research has also focused on:

-improvement of the combustion efficiency, through a better design of the combustion chamber and an improved location of the injectors;
-increase of the average operating pressure;
-increase of the operating temperatures;
-improvement of thermal insulation (development of adiabatic motors);
-reduction of friction resistance.

Fig.11

Fuel injection pump controlled by an electronic system
(Source: Bosch)

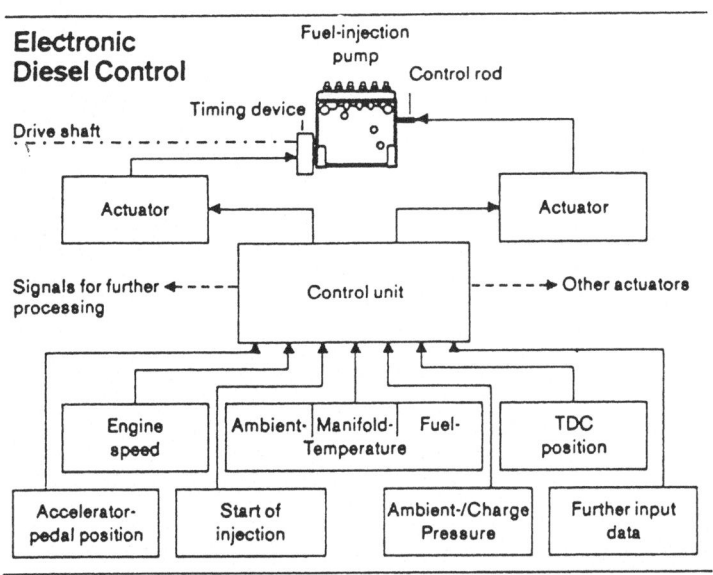

Considerable progress has been achieved for the two first points and new developments are expected to continue. With regard to the friction reduction problems, table 12 gives the mechanical losses typically measured in a diesel engine.

table 12

Mechanical losses in a typical diesel engine for a tractor
(Source: Clowing, 1984)

Friction source	Power loss (kW)
Pumping loss	7.9
Piston/rings	4.1
Crankshafts/bearings	1.3
Fuel injection pump	0.7
Water pump	0.5
Oil pump	0.4

There are not any specific areas where substantial improvements in efficiency can be attained. Nevertheless, a careful tuning could yield, without excessive cost, to further energy savings of around 2-3%.

With regard to the possible improvements in heat insulation, the aim is to reduce the 30% energy loss that occurs through the cooling and lubricating systems.

Good results could be obtained by the use of ceramic materials, which allow to rise the combustion chamber temperature, enlarging the area of the thermodynamic cycle and leaving a higher energy content in the exhaust gas, and so making possible further improvements in the turbo-charging techniques.

Another possibility is to couple the turbine driven by the exhaust gas with the crankshaft, giving rise to the so-called "turbo-compound engines". The development of this technology requires considerable modifications in the structure of the engine and of its components. This is a long term development, leading to efficiency improvements of around 5-7%.

The use of ceramic materials will only be achieved by means of heavy investments: this development therefore will only affect agricultural engines after ceramic materials have been introduced in road vehicles.

As a conclusion, we summarize as follows the trend in the development of diesel engines:

1)-increased use of electronic control;

2)-increase of the cylinder pressure;

3)-reduction of the weight of the engines, the power output being equal;

4)-reduction of the specific consumption.

Fig. 12 and 13 show the first and the third points of this summary.

Fig.12

Electronic controller for commercial diesel engines for the year 2000
(Source: Millar,1985)

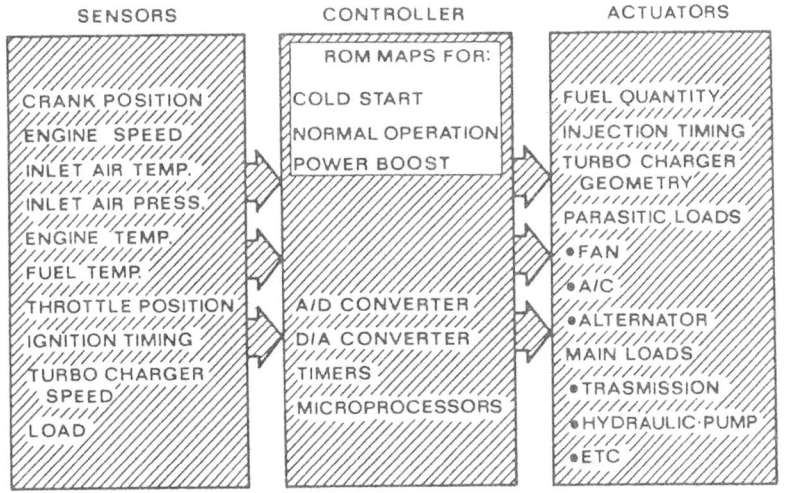

Fig.13

Diesel development trends
(Source: Millar, 1985)

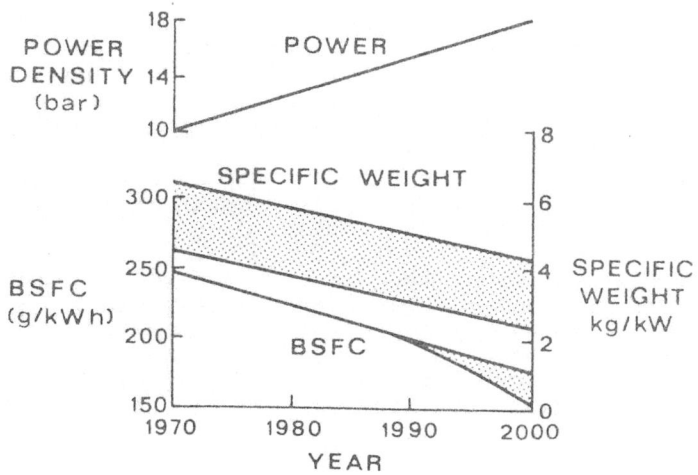

Finally, a brief mention must be made about the necessity to consider the production of agricultural engines in the light of the results recently introduced in road vehicle engines. For example, in the Fiat's engine "Fire" the number of components in the engine has been reduced by 30%.

We can define the fields in which, in the short-medium term, efficiency improvements will be possible for agricultural engines:

-design improvement and use of new materials (4-5% gain);
-turbo-charging in low power engines.

In the longer term, the use of ceramic materials could represent a new way forward for all whole engine sector, bringing further reductions in fuel requirements.

3.1.2 - Tyres

The internal losses, the mechanical energy developed by the tractor engine is transmitted by wheels complete with tyres. Their function is to provide the necessary adherence for the motion of the tractor itself, overcoming the frictional resistance and delivering a tractive force for the drawn implements.

The coupling between tractor and the implements may be different for the implements mounted or drawn by tractor. Consequently, power supplied to the implements may vary (only traction, just the couple to the power take-off, pressure for hydraulic systems etc). In some cases the traction power prevails, in other cases the power to overcome the frictional resistance is insufficient.

Traditionally, in the analysis of tractor tyre mechanics, the tractive function has always been considered the most important. The result has been the improvement of the adherence coefficient by means of the structure of the tyre itself, its flexibility, and -for the driving wheels- by the tread design. The adherence coefficient is presently in the range from 0.2 (seed-bed and highly deformed soil) and 0.6 (mud floor road). The rolling resistance is rarely considered, even though this parameter is important both for driving and the front wheels, particularly if it reaches very high values on loose soils.

The trend over the last 8-10 years, has been to introduce radial ply tyres and, very recently, radial high flexibility models. The use of new models can give the results illustrated in fig.14: on average (at 15% slip), the useful tractive power increases from 1 (conventional tyres) to 1.15 (radial ply tyres) and to 1.30 (radial ply with enlarged cross section flexible tyres).

<u>Fig.14</u>

<u>Comparison between conventional, radial and radial large
tractor tyres performances</u>
(Source: Gasparetto, 1984)

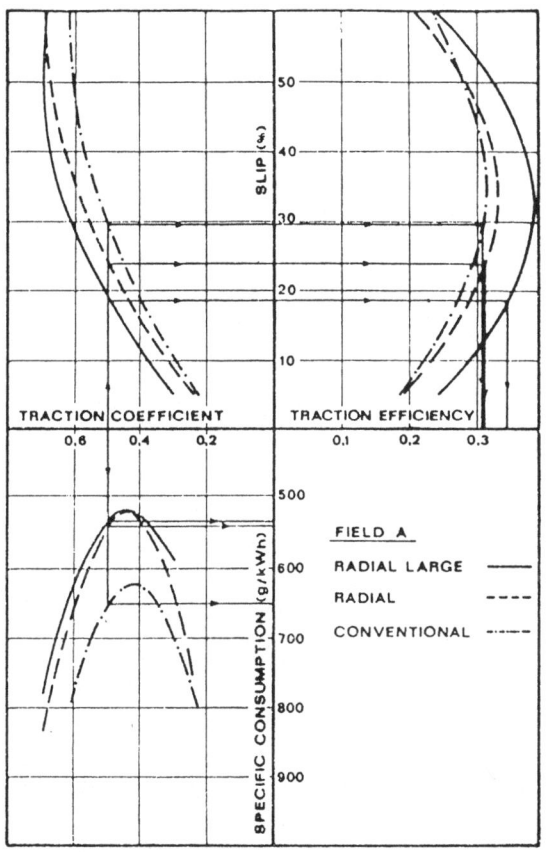

The above mentioned performances represent, under the
same conditions of power supplied, a 20-25% improvement of
the tractive power or a 15% improvement of the global
efficiency of the tractor.

In spite of the improvements already attained, the
widespread use of flexible tyres with a good floating index
is still a long way off. This type of tyre with a reduced
strain and low soil compaction, could give the following
advantages:

-to reduce the direct energy losses by 15-20%. This figure corresponds, other parameters being equal, to a fuel saving of 3-4%. This is valid for both the driving and the steering wheels.

-to make work possible, without serious damage, to soils with a low carrying capacity. This means an increase of the utilization time and a reduction of the nominal power required.

-to reduce damages to the crops and thus indirectly resulting in higher yields.

Finally, in the future, a high performance tyre may result in a direct energy saving of 8-10%. Further indirect savings of 3 to 7% must be added, according to the quality of soil and the type of crop.

3.1.3 - Electronic control devices

The development in the use of tractors, has led the manufacturers to equip them with control systems in order to have a continuous monitoring of the various components.

Fig.15

The elements of a microprocessor-based monitoring and control system
(Source: NIAE, 1983)

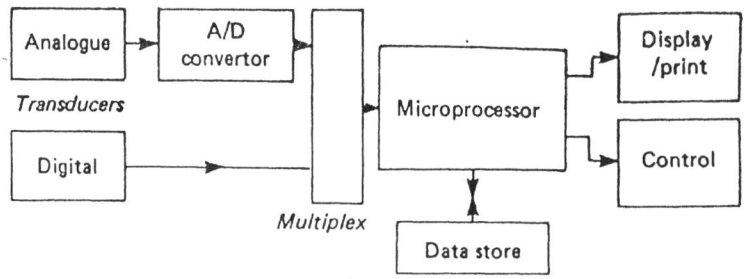

This kind of system has been developed because of the sophistication of modern farming and also because the driver is commonly enclosed in a cab and is therefore not in close contact with the operating components and he needs at least some indirect control.

Beside the instruments which have been used for many years, other new instruments have been added to improve the performances of the tractor: oil pressure in the hydraulic system indicator, air filter cloggage indicator, on-off duty indicators for power take-off, for differential gear lock, for front-wheel drive, for parking brake, etc. All those controls lead to better operating conditions and, finally, to some energy savings.

However, other possibilities of electronic control systems do exist. Control systems (Fig.15) could make a significant contribution to energy saving in three ways:

-supplying the monitoring of various parameters;

-using micro-processors which, through the comparison of different data, provide the driver with useful information for optimizing the use of the machine;

-using automatic control devices to keep in the optimum range of the operating conditions.

Some manufacturers have already introduced this type of monitoring and many research institutes (as do also the manufacturing companies) are engaged in the development of more complete control systems.

Renault has marketed a device called "Eco-control"(Fig.16) based on the simultaneous monitoring of the engine revolutions and the exhaust gas temperature. When the needles of the two indicators are both in a given region of the dial, the driver knows the vehicle is operating at optimum conditions in terms of fuel consumption. The system is based on the correlation existing between the exhaust gas temperature and the engine load.

The same manufacturer is expected to market, in 1986, a new and more sophisticated control system called ACET (Aid to Economic Tractor Driving) able to lead to savings of around 12%.

Fig.16

Eco-control: a dash-board instrument mounted on some Renault tractors

(1: gear scale; 2: rev. counter needle; 3: rev. counter scale; 4: temperature (engine load) indicator scale; 5: temperature (engine load) indicator needle; 6: hours accumulator; 7: black area; 8: green area)
(Source: Renault)

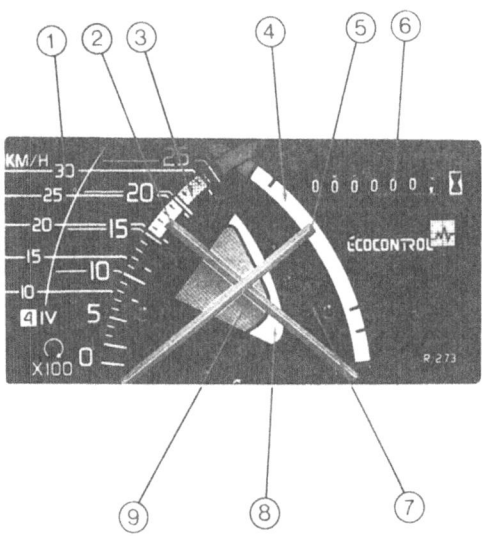

Pang et Al., of the Saskatchewan University (Canada) suggested the direct monitoring of the exhaust gas temperature as a way of evaluating fuel consumption. As shown in Fig.17 the average error is less than 2%.

One of the most interesting aspects of this system is the elimination of the traditional expensive flowmeters. For use in field operations this instrument is not costly with respect to its performance.

Laib and Komandi of the Gödöllö University (Hungary) also developed an instrument based on the exhaust gas temperature. This instrument records, on the basis of the time and the gas temperature, the "history" of the work done by the tractor, supplying useful indications for the farm management.

49

Fig. 17

Monitor performances measured by controlling the exaust
gases temperature

(Source: Pang, 1985)

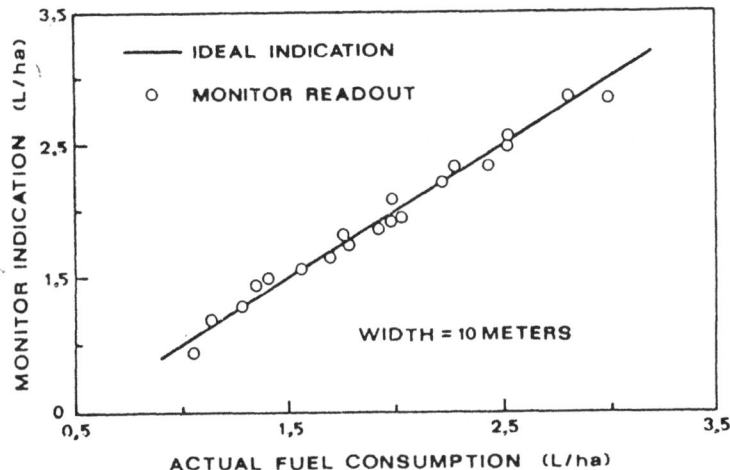

The important limitation of the information supplied by
this type of instrument, is that the driver does not know
the slip value even though, in the tractive operations,
knowledge of this parameter is essential in order to get
optimum performance. This is the reason why many researchers
are developing indicators that monitor the difference
between the real advancement speed and the speed
corresponding to the driving wheels. To overcome this
problem, two solutions have been considered.

The first solution is to install aboard a small radar
able to monitor the real working speed by means of the
Doppler's effect (Fig.18).

Fig.18

Ground speed measured by radar system
(Source: Renaud, 1984)

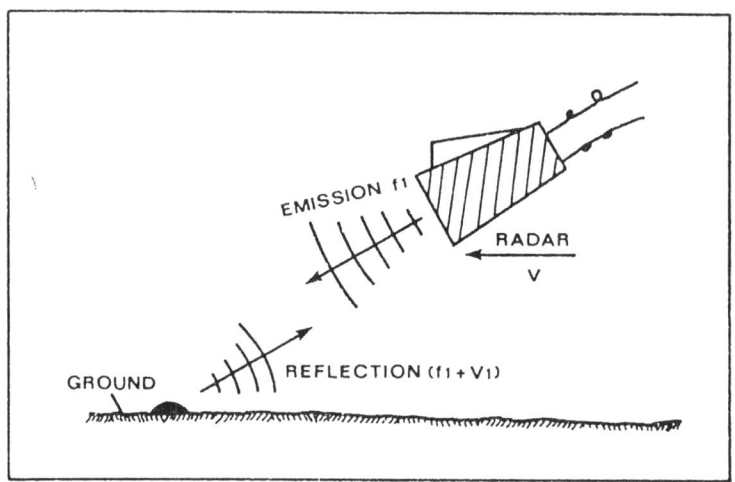

Richardson et Al., recorded errors due to the irregular oscillations of the tractor versus the ideal advancement trajectory. Nevertheless this solution, tested in different conditions, gave errors of only 3% when the integration time allowed for each recorded point was 1 second. On the base of this type of instrument, complex monitoring systems have been developed which give the driver the necessary information for an economic and optimum utilization of the tractor.

Fig. 19 shows the monitoring scheme applied by NIAE (Silsoe). This system uses measuring devices connected within the dash-board in a conventional manner. On the dash-board itself it is possible to read the measured data both in alpha-numeric form or as graphs. Variable parameters can be introduced in the system, like the working width or the soil surface conditions. This system does not give the driver any direct information on how to improve the tractor performances.

Many tractors and electronic component manufacturers have begun to offer monitoring systems based on the measurement of the real speed.

Fig.19

Diagram of NIAE tractor performance monitor
(Source: Gasparetto, 1984)

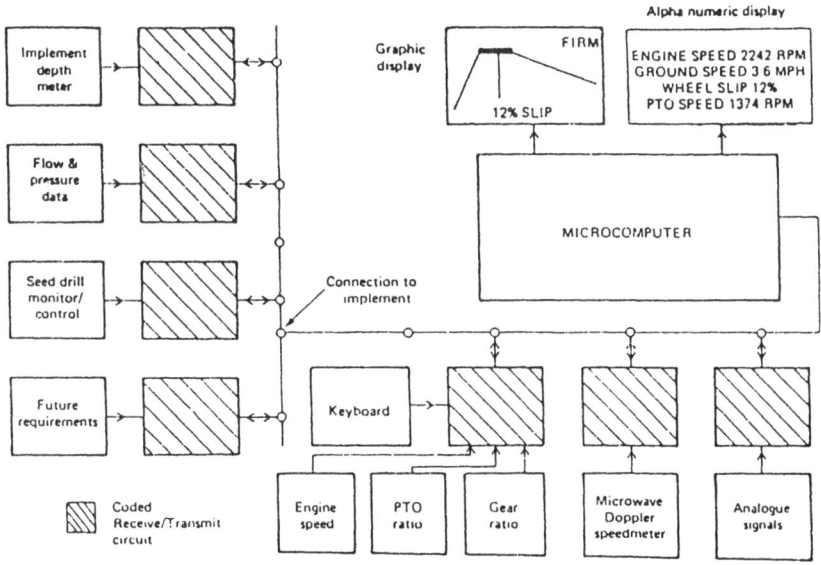

Beside the instruments already described, in recent years many automatic control systems have been investigated. A common feature of this type of control system is the use of a microprocessor analyzing the data supplied by the various sensors and making decisions which are put into effect by the system itself.

One of the simplest examples is for the control of the hydraulic lift. These are standard devices mounted on high power tractors and their use improves the ease of operations, the precision drive and the response time. One of these systems is based on sensors which monitor the inductance variations in a coil when the position of an iron rod changes (fig.20). These variations are processed by a computer which, by means of an actuator, controls the oil pressure of the hydraulic system modifying the lift position.

Fig.20

Electronic hitch control for tractors
(Source: Renaud, 1984)

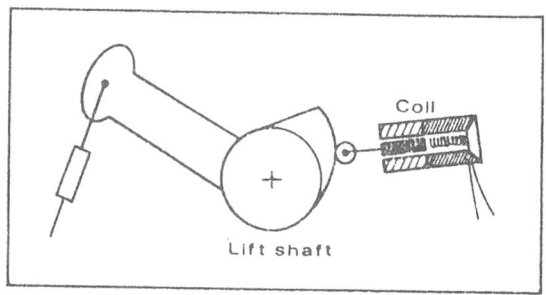

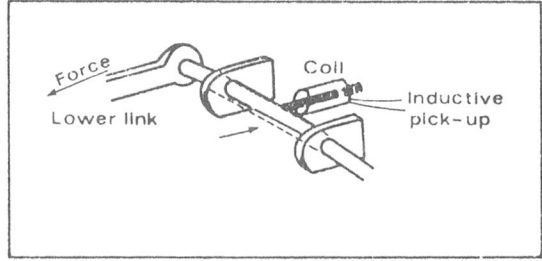

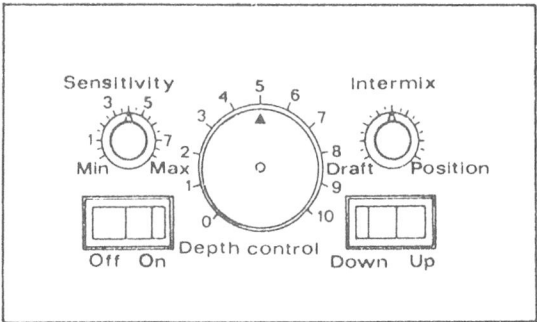

The choice, as in traditional systems, is to carry out
the control either by position check or by power check. A
mixed type of control is possible by means of knob drives on
the dash-board. The system may be integrated by the
information of the reaction speed of the lift.

Following Dweyer, the improvement in the performance of
the systems based on the power check, can lead to a
reduction of about 10% of the traction effort.

We point out that the improvement of the lift control
efficiency has the additional advantage in that the tractor
is running most of the time close to its maximum power,

thus giving a better performance and greater fuel economy.

Other automatic control devices that have been marketed by the manufacturers include automatic differential locking and the coupling of the front driving wheels.

The John Deere Co. offers a line of tractors in which the front drive coupling is automatically operated when the exhaust gas temperature exceeds a pre-established value. The limitation of this system is that it also actuates the 4 driving wheels when the engine is supplying its maximum power to the power take-off and so the power supplied to the implement decreases and the power available for the advancement increases needlessly. The plug-in of the 4 wheel drive should be controlled by the slip.

An Italian manufacturer (SICE) has investigated an electronic device for controlling automatic differential locking (Fig.21).

<u>Fig.21</u>

<u>Automatic differential locking</u>
(Source: Renaud, 1984)

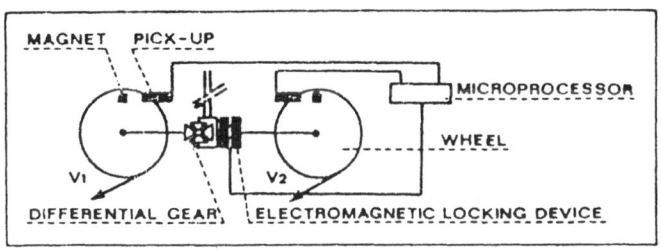

Both the front wheels are equipped with speedometers sending a signal to a computer: this calculates the difference in the speed of the two wheels and, if this difference exceeds a pre-established value, the differential is automatically locked.

For several years the automatic control of the transmission ratio has been considered to be a valuable objective for the optimum utilisation of tractors.

In high powered tractors there are other possibilities of automatic control, e.g., mounting epicycloidal gear boxes

gear boxes and shifting the various gear ratios by an
hydraulic drive (Fig.22).

Fig.22

Electro-hydraulic control of tractor transmission ratio
(Source: Renaud, 1984)

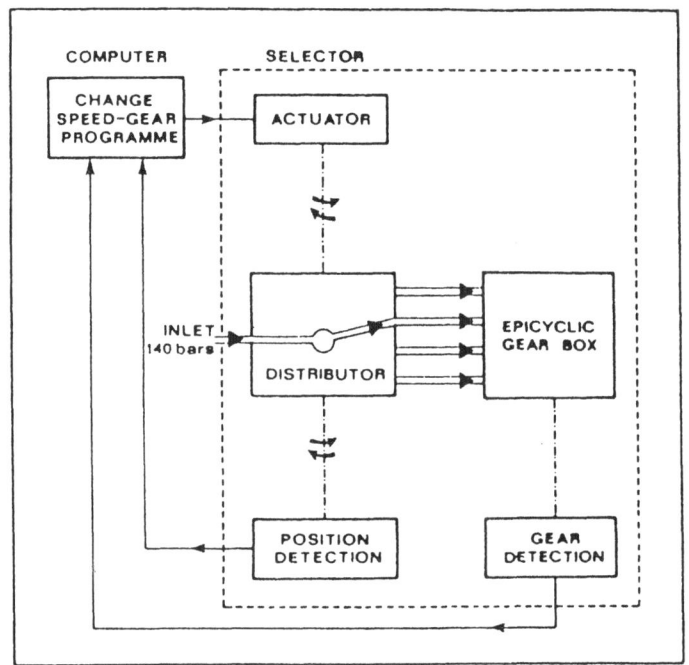

In this system, the hydraulic distributor orientates the
high pressure oil flow by means of electric valves, either
braking or releasing some elements of the epicyclic gear. A
microprocessor correlates the information about the real
speed and the engine load. On this basis, the opening and
the shut-off of the electro-valves is controlled and an
automatic gearshift is obtained.

Another automatic gearshift system that can be installed
on traditional gear transmissions has been recently proposed
by Chancellor and Thai (Fig.23).

Fig.23

Electro-mechanical control of transmission ratio
(Source: Chancellor, 1984)

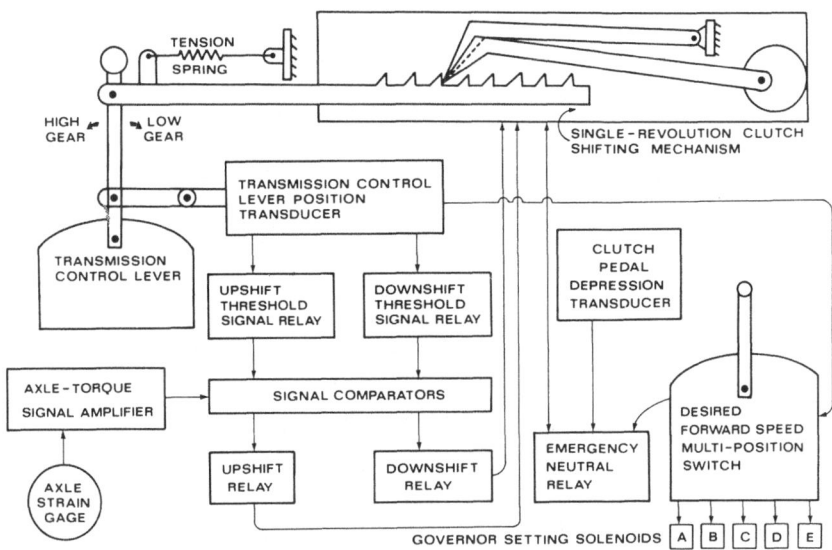

This system is based on the control of the transmission rate and of the engine r.p.m. by means of strain-gauges measuring the engine torque. The driver chooses the advancement speed he wants and sets the value directly on the dash-board. A computerized program actuates the gear ratio on the basis of the fixed speed and of the strain-gauge information: the choosen gear ratio corresponds to the minimum fuel consumption. A prototype of this system has lead, in field ploughing and harrowing, to 5-12% fuel savings, the working capacity being simultaneously increased. The important limitation of the system is the lack of slip monitoring, however there are no devices commercially available, until now, that integrate the engine torque and slip for the gear ratio control.

Automatic steering is an other point which is becoming the subject of investigation in many countries. Following Young et Al., this automation is important because in many cases the required driving precision is beyond the capacity of the driver, so that the machine is not utilized to its maximum performance (Fig.24).

Fig.24

<u>Leader cable automatic guidance system: block diagram</u>
(Source: Young, 1983)

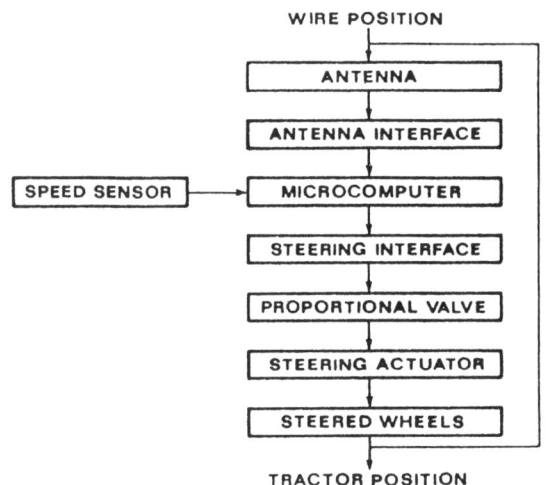

The introduction of automatic steering systems should enable energy savings through:

-working capacity increase and
-increase of the crop yields due to less soil compaction.

The last sector of the research on automation is the continuous control of the tyres inflation pressure. Some Scottish researchers have examined this possibility, in a pre-established range, as a function of the slip. The idea is to lower the inflation pressure when the slip increases in order to increase the adhesion, and vice-versa. Therefore an air compressor must be installed on the tractor and also a complex device is necessary to inflate the tyre when in motion. It is unlikely that this system will generate much interest in the near future, considering the long response time inherent in this kind of device.

It seems that in the coming years automation and automatic control are capable of introducing improvements in tractor efficiency. These improvements can be translated into energy savings: according to the type of the soil, the operations requirements and the size of the fields. The range of these savings will be between 7 and 20% of the present energy requirements.

3.1.4 - Tractor architecture

Farm and extra-farm transportations may not require high traction forces, as speed becomes increasingly important. Specific requirements are occurring in some production sectors like horticulture and mountain-farming.

To meet so many scattered demands, the manufacturers answer is to offer specific machines, planned with a modified structure in order to enhance a given operating performance. So the following have been introduced:

- 4 wheel-drive tractors with the four wheels of the same diameter, in order to obtain high potency in tractive operations;

- low barycenter tractors, specially conceived to operate in mountainous farms with easy handling;

- tool-carriers and vehicles for special operations.

Considering the diversity in the farming sector needs, it is difficult to sketch out the evolution of tractor architecture. The trend is to consider the tractor as a "travelling power station" even though, in the tractors mass-production, no modification appear in the classic architecture. This is still based on a standard rigid structure, on which the driver sits straddle-legged.

The implements structure is, on the other hand, based on the production of (front- or rear-) mounted models, with operating elements powered by the power take-off, electrically, or hydraulically. Nevertheless, this is not true for ploughing and transportation, which only requires tractive forces.

Therefore the tractor architecture is being examined again, as did the NIAE in its prototype (Fig.25).

58

Fig.25

Experimental NIAE tractor able to operate with different
implements and for several operations
(Source: Febo, 1985)

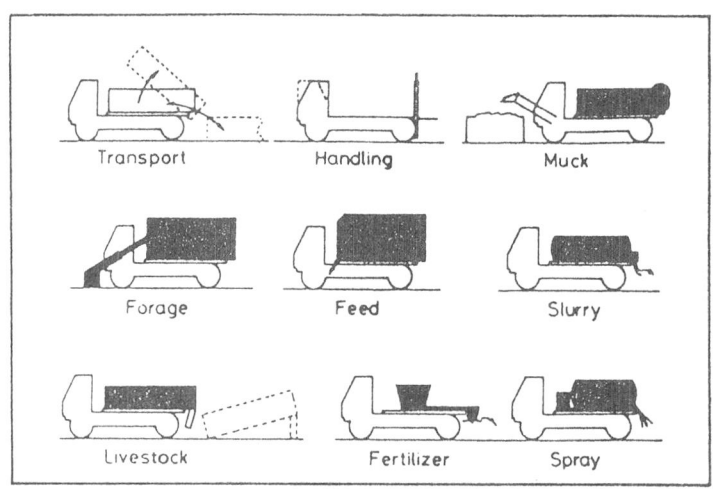

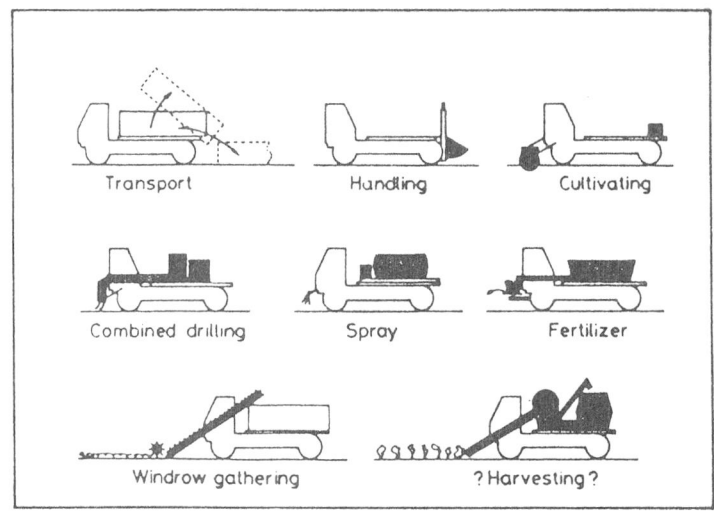

Two different versions are proposed:

- One version is of the "travelling power station type" with carriaged tools, 4 driving wheels of the same diameter, structured as a transporter, mainly for internal farm use but suitable also for transportation with a maximum speed higher than 40 Km/h;

- The other version is of the present tractor type, mainly assigned to utility enterprises, having high power and suitable for heavy towing operations.

This type of development (still to be confirmed) could result in a significant reduction in fuel consumption because of greater utilisation, a better fit between the tractor and the operating machines and finally, to the reduced energy losses.

Farmers are demanding higher work speeds than are presently found, especially during the transportation operations. To meet this requirement the dynamic behaviour of the tractor must be modified in order to improve the driver's comfort and to reduce the vibrations by means of suitable suspension.

This improvement has already been adopted by Mercedes Benz (Fig.26) and by some manufacturers of tractors specifically designed for mountain duty. Probably this trend will spread in the future, because it allows a more flexible planning and, therefore, a higher diversification of the machines.

As a complement to this type of solution, it is interesting to recall Matthews' forecast of a revolution in tractor utilization. Because of the increase of farming specialisation and the decrease in price of electrical energy, the future operating machines will be made up as tool-carrying gantry frames, electrically powered and working on wide operation strips (Fig.27). The wheels of the gantry will run on rails preventing soil compaction and reducing the unproductive surface.

The gantries currently in use are mostly for pesticides and fertilizers distribution but they could also be used for tillage and for seed-bed preparation.

The power required for this type of operation may be only 10% of the power presently installed on tractors and could be entirely supplied as electric energy, with the consequent obvious energy savings.

In this system all the operations could be highly automated (also the driving control) by means of electronic devices and for some operations, like phytotreatments and fertilization, the machine could work without operators.

Fig.26

Frame-type Mercedes Benz tractor with 2-wheels suspension
system
(Source: Mercedes Benz)

Fig.27

NIAE gantry system
(Source: Mattews, 1982)

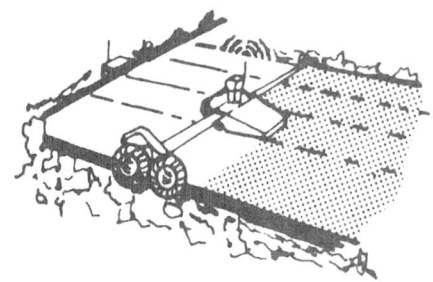 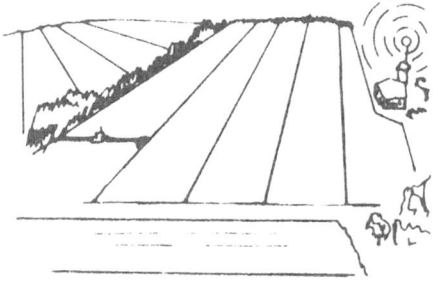

3.1.5 - Fuel oil replacement

The continuous fluctuation in the price of oil derived products and the necessity to achieve higher efficiency levels in tractor management, in recent years has brought to the foreground the investigations into alternative fuels.

Prototypes of farming machines have been proposed which are based on the use of:

-biogas obtained from anaerobic fermentations;
-air gas from thermochemical gasification processes;
-ethanol and methanol;
-vegetable oils;
-electric power.

The problems of using biogas in tractors are related to the necessity to pressurize it up to 200 bar, so that sufficient quantities can be carried. The hydrogen sulfide usually present must be eliminated due to its corrosive action. For this purpose, downstream of the production plant, a very complex and expensive purification-compression plant is necessary. Once purified and compressed, the biogas does'nt create technical problems: a dual-fuel technique is adopted using a small proportion of fuel oil (Fig.28).

Fig.28

Example of a diesel-engine transformation in a dual-fuel engine: 1, biogas tank; 2, valve; 3, high pressure reductor; 4, safety valve; 5, low pressure reductor; 6, actuator; 7, proportioning device; 8, limiter; 9,10, injection pump; 11, accelerator; 12, potentiometer.
(Source: De Zanche, 1983)

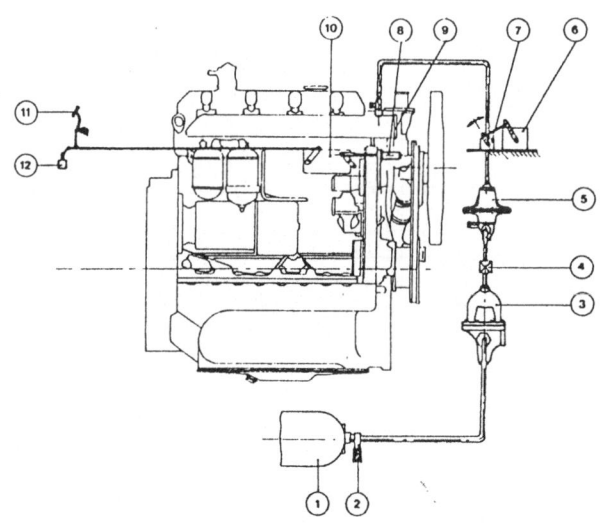

Following the De Zanche's experience, comparing the mixed cycle and the diesel cycle, the same maximum power value is reached but:

1) 450 rpm are lost at the maximum power rate, probably due to the poor amount of air introduced (Fig.29);

2) the engine stalls at low speed (around 1.150 rpm), due to an eccessive gas feed (Fig.29);

3) at the maximum power rate the fuel oil consumption is between 24 and 28% of the total.

Fig.29

Comparative performances of an engine fed by diesel fuel and dual-fuel (diesel fuel and biogas)
(Source: De Zanche 1983)

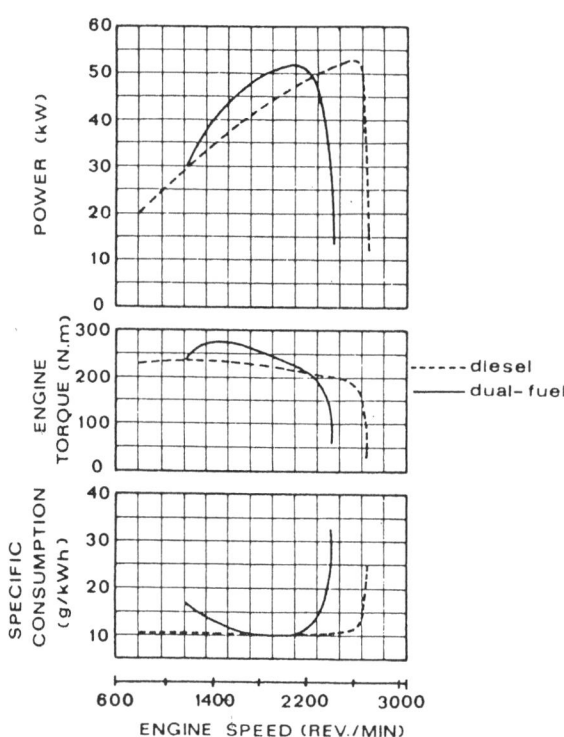

63

Thus, the biogas replaces 50-70% of the fuel oil consumption. Swiss investigators are also working on this technique. Its main limitation is the high cost of the gas production and its epuration and compression. Even though results are promising, it is impossible to predict, in the short-medium range, the practical utilisation of this solution.

In the field of the air low energy gas, presently its use appears limited to fix point utilizations. There are also many uncertainities from the economic aspect. However, in recent years, particularly in France and Belgium, many researchers have been investigating the feasibility of using gasifiers mounted directly on the tractors(Fig.30).

Fig.30

Tractor powered by a diesel-low energy gas engine
The gas is produced by charcoal on the tractor
(Source: CEMAGREF, 1986)

Several solutions have been advanced: the use of low energy gas coming from ligno-cellulosic biomasses, without a previous transformation into charcoal and feeding Otto cycle engines; the use of low energy gas obtained from charcoal to feed diesel dual-fuel engines; and so on. On average, the power obtained is -for the first solution- only 40% of the nominal power (Fig.31). For the diesel dual-fuel engines, when 10-15% of the energy is supplied as diesel oil, the power obtained is 60% of the nominal power.

Fig.31

Power curves of the same engine fed by gasoline and
low-energy gas
(Source: Vaing, 1983)

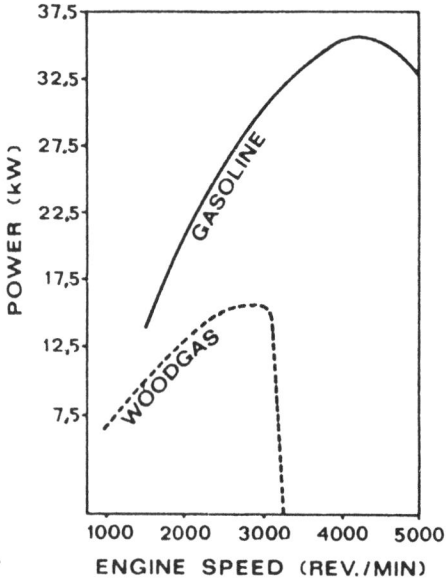

The thermodynamic efficiencies are around 15-20% and, therefore, they are much lower than with the traditional fuels.

When the use of non carbonized biomass is proposed, the problem is the obtention of a sufficiently purified gas. This requirement necessitates the use of expensive filtering apparatus, the maintenace costs of which are also very high. The use of charcoal is easier, but its production is also expensive, both from the economic and the energy point of view.

In the short term, there are few opportunities for the development of this technique in the developed countries. This may not be true for the developing countries.

Recently the production of <u>ethanol</u> from agricultural by-products has been discussed. The use of "energy crops" and the utilization of agricultural surpluses have been suggested too. Considering the complexity of the process, the economic size of the production plants is large. This kind of activity must be considered as an industrial business, requiring large financial and technological investments. Therefore, the ethanol will probably be put on the market via the existing distribution nets of oil derived products. Though, looking at the physical characteristics of some fuels gathered in <u>table 13</u>, it clearly appears that ethanol is closer to gasoline than to diesel oil.

<u>table 13</u>

<u>Main physical properties of different fuels</u>

Properties	Units	Ethanol	Gasoline	Diesel-oil
Net heat value	kJ/kg	26.940	42.690	42.850
Evaporation temp.	°C	78	30-190	180-360
Evaporation heat	kJ/kg	854	289	252
Octane number	-	106	97-99	40
Cetane Index	-	5	10	50
Stecheometric ratio	-	8.95	14.4	14.4

In the mixture ethanol/gasoline, ethanol acts as an antiknocking agent and thus, can replace the lead compounds, with evident advantages from the environment point of view. In the mixtures ethanol/diesel-oil many problems arise, particularly at low temperatures where a phase separation takes place. A second problem comes from the gas-plugs which are formed because of the low boiling point of ethanol.

If ethanol has to be used in diesel engines, the dual-fuel system must be preferred. Obviously, this may result in engineering complications and poor performances.

It may be more advantageous to investigate the use of special engines with a compression ratio as high as 12:1, a special injection pump and sparking-plug ignition.

Finally, if the ethanol as fuel were introduced, the car sector will be more important than farm tractors.

As far as <u>vegetable oils</u> (rapeseeds oil, etc.) are concerned the situation is about the same. Engines suitable for a feed based on vegetable oils have been developed, which give outstanding performances and thermodynamic efficiencies up to 40% (Elsbett). Positive experimental results have been obtained by Pernkopf indicating that the behaviour of vegetable oils is similar to diesel-oil when

used either in precombustion engines or in direct injection
engines, with similar performances (Fig. 32).

<u>Fig.32</u>

<u>Performances of the same engine fed by: A) 100% diesel-oil;</u>
<u>B) 50% colza oil + 50% diesel oil</u>
(Source: Pernkopf, 1980)

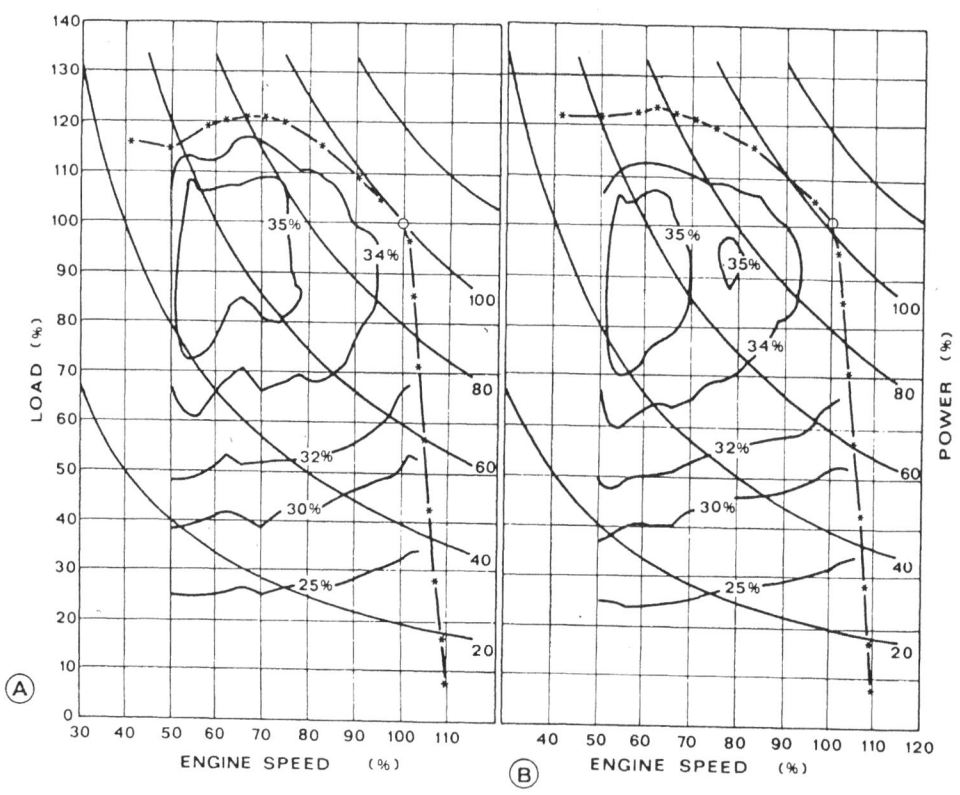

Pernkopf point out that vegetable oils are generally more viscous than diesel oil, so that they must be blended with diesel oil and pumped using in-line injection pumps. Furthermore, not all the vegetable oils can be used as fuel. Linseed oil, for example, is not recommended. An important aspect of the use of vegetable oils as fuel is the low cost of the modifications necessary to adapt the existing diesel engines, when this solution is compared with the other possibilities, as in table 14.

table 14

Replacement possibilities of diesel-oil for tractors

Solutions	Investments (% of tractor cost)	Diesel-oil replacement (%)	Cost of fuel vs. diesel-oil (%)	Remarks
Gasoline + diesel-oil	none	max. 30	+5	safety
Ethanol + diesel oil	none	max. 5	+5	stability of mixture
Vegetal oil + gasoline	low	100	+80	safety,resin
Vegetal oil + diesel oil	low	50-100	+50-100	viscosity
Dual fuel(diesel + ethanol)	30	max. 50	+150	no experiences
Dual fuel(diesel + wood gas)	70	max. 80	-	weight; safety
Ethanol	70	100	+350	spec.reservoir
Liquid gas	90	100	-25	safety; weight
Wood gas	120	100	-	weight,safety power loss

(Source:elaboration from J. Pernkopf, 1980)

Also in this case, the greatest potential seems to exist in the motor vehicles sector: in fact, the main supporters of this technique consider the automobile market to be more important.

Finally, the use of underline electric energy in the farming

68

activities appears to have some potential. Farming
activities are characterized by a power demand which varies
with time. A recent analysis made by Dweyer gives the torque
requirement distribution in ordinary forage harvesting
operations when the power take-off is used.

<u>Fig.33</u>

<u>Torque requirement distribution in forage harvesting</u>
<u>operations</u>
(Source: Dweyer, 1984)

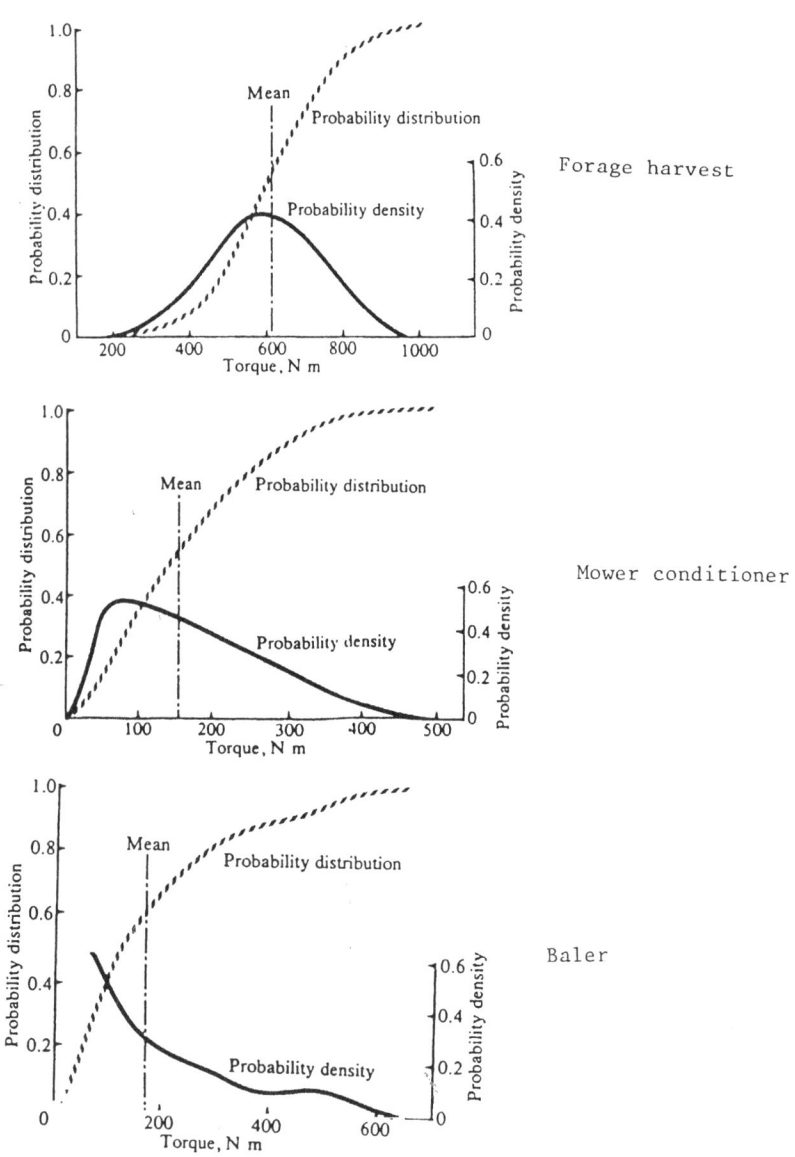

As shown in <u>Fig.33</u>, the energy requirements are variable
in this type of operation. Using a diesel engine the motor
works far from its optimum conditions and its efficiency in
transforming the energy of the fuel is, on average, around
25-30%. Moreover, during the harvesting operations, often
the engine is running idle: even if the operating machines
are properly coupled with the tractor, the diesel engine
efficiency remains poor.

The main characteristic of electric engines powered by
direct current is that they have a high efficiency (0.8-0.9)
even though the torque requirements vary continually.
Moreover, over short periods, this type of engine can supply
torques exceeding the values corresponding, at the same rpm,
to the average power of the engine itself. The performances
of a tractor powered by a direct current electric engine are
certainly superior to that of a diesel engine of the same
power. The high efficiency allows the size of the engines
and the capacity of the accumulators to be reduced.

One of the first electric tractor prototypes built in
Italy with a nominal power of 20 kW has given performances
equivalent to a conventional 45-50 kW tractor (<u>Fig.34</u>).

<u>Fig.34</u>

<u>Scheme and dimensions of the electric tractor</u>
(Source:Gervasio, 1984)

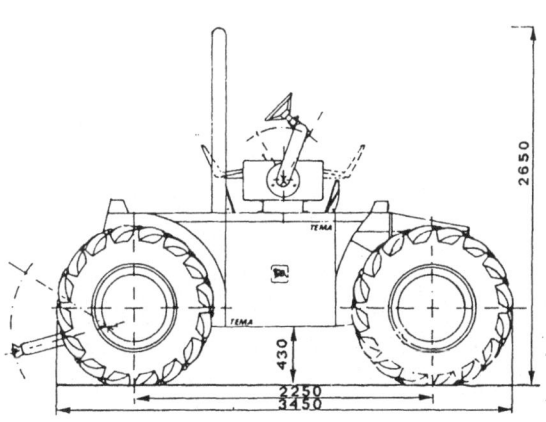
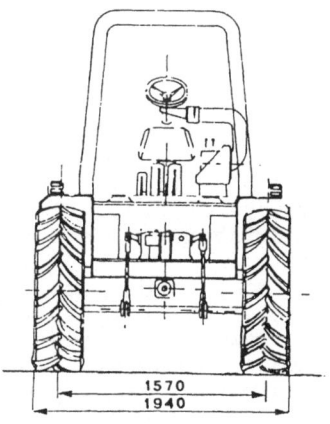

The manufacturing process of the electric tractor powered by a battery is simpler than the diesel one. <u>Fig 35</u> shows its greater simplicity. A mechanical gear box, in medium-small size tractors, has 12-20 forward gear ratios and 6-10 reverse gear ratios: in the electric tractor the gear box is replaced by an electric control interposed between the battery and the electric motor, which regulates both the voltage and the current flowing to the motor as operation requirements change. This is possible because the electric motor can be used in ranges of speed and torque much wider than the internal combustion engine.

<u>Fig.35</u>

<u>Scheme for the propulsion of diesel and electric tractors</u>
(Source: Gervasio, 1984)

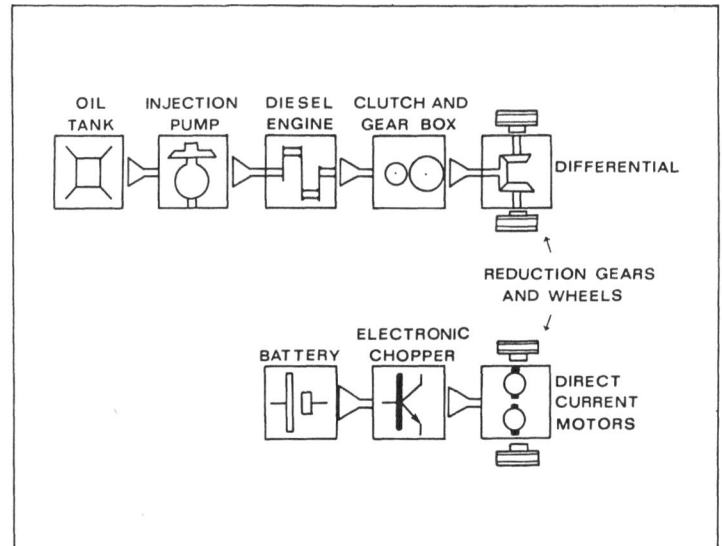

The reliability of electric motors and transistorized converters presently offered on the market is well established.

Today the most important technical limitation of electric tractors is the low capacity of the lead batteries. The lead batteries offered on the market have a limited charge density and low efficiency in the charging-discharging cycle (60-70%).

In <u>table 15</u>, drawn from a recent Alcock study, the energy density of some fuels ordinarily used in agriculture is compared with two different types of batteries.

table 15

Comparison of fuel energy densities

Fuel source	Energy density (MJ/kg)	Conversion efficiency assumed	Fuel mass equivalent to 100 liters of petrol (kg)
Petroleum	44.2	0.20	74.1
Diesel-oil	43.0	0.26	58.5
Lead-acid battery	0.086	0.504 *	15,112
Advanced lead battery	0.135	0.504 *	5,656-9,627

* A battery charge-discharge efficiency of 0.7, a controller efficiency of 0.9 and an electric motor efficiency of 0.8 were assumed. These are cited as typical efficiencies by the respective manufacturers, viz: -General Battery Corporation, Pennsylvania: Cableform. Stockport. Cheshire: and General Electric Company. Eric. Pennsylvania.
(Source: Alcock, 1984)

In spite of the higher conservation efficiency of the electric systems, a large mass of batteries is necessary for vehicles of equal performance. In table 15 the production and transportation efficiency of electric power is not considered. This efficiency is around 0.4, thus the global efficiency for the transformation of the primary energy into mechanical energy, for both the electric and the fuel based systems, is of the same order of magnitude.

Therefore, the advanced lead batteries does not lead to any significant energy savings. However, an improved utilization and distribution of the electric energy could be achieved if the batteries were recharged by night, when the electric power is supplied at a reduced rate.

This type of technique could be developed only through the use of new types of batteries (nickel, lithium, cadmium etc.) having higher energy densities in order to avoid overloading of the vehicles and to increase their utilization.

In some cases (stables, green-houses, etc.) the electric energy can be supplied by cables. Alternative, hybrid systems do exist, using electric and thermal propulsions combined into two basic schemes: one with series connection, the other with parallel connection. These systems have not yet been used in agriculture, but are capable of good results, in terms of energy savings, as indicated by the results already obtained for automobiles and dumpers.

3.1.6 - Operative conclusions

Among the different innovations and solutions which have been suggested, some may be successful in the medium range; others only in the long range, depending on the general development of motor traction; others do not seem to have any chance of becoming operational.

The use of microcomputers at the engine level and, in a more general way, for the optimization of different operations; the use of radial ply tyres with enlarged cross section; the use of turbo-charged engines in small tractors are all developments which will require at least ten years before being wide spread in farming operations: however energy savings of around 12-15% of present consumption should be obtained. This means, for the EEC, a saving of around 1.4-1.6 Mtoe/year, but this figure could be in excess of 2 Mtoe/year, when Greece, Portugal and Spain reach a higher level of mechanization.

In the long term, the introduction of new materials (particularly the ceramic materials), the modification of the tractor design and the use of "electric solutions" may bring an other 10% of savings, that is 1.1-1.2 Mtoe/year. Incentives will be necessary in this field.

The possibility of successfully introducing alternative fuels for tractors, even if these fuels come from renewable sources, is more likely to occur in developing countries.

In conclusion, by the year 2000, a total energy saving equivalent to 2 Mtoe/year would seem to be possible. This result could be attained only if the necessary incentives are forwarded and put into action.

3.2 - Primary and secondary tillage machines

Thirty eigth percent of the direct energy requirements for field operations can be ascribed to soil tillage. This is the sector where there are large energy saving opportunities. Of special interest is the research carried out by Patterson at the NIAE. His results can be summarized as follows:

- the primary tillage operations require 75% of the total energy spent before the seed-time;

- a significant proportion of this energy (up to 50%) can be saved by the use of implements other than the conventional plough;

- a further decrease in energy consumption can be obtained through the practice of sod-seeding: this technique uses only a fraction (12-22%) of the energy required in comparison to traditional techiques;

- considering the nominal power required for the tractor, the greatest advantages are obtained by the use of tools mounted on the tractor itself and powered by the power take-off. However, the use of rotating tools is limited (for agronomic reasons) in some European lands.

- for the same tillage depth, the tools with tined implements require less energy than tools with mouldboard implements. In fact, using mouldboard ploughs, a proportion of energy is used to cut the soil, an other fraction is used to overcome the friction between the soil and the tool, and finally energy is needed to turn the slice upside down. The energy needed to overcome friction can represent up to 80% of the total energy requirements.

Many research programs have attempted to reduce the soil/tool friction, aiming to replace the traditional methods by alternative systems. Some of these programs are the continuation of early research proposals, already advanced at the beginning of the century.

3.2.1 - Ploughs

The considerations summarized above have lead, in recent years, to the development of mouldboard ploughs having a structure that results in less energy absorption. Open mouldboard ploughs have been adopted (Fig.36) and new designs that allow a lower lifting and easier overturning of the soil slice (diamond-type mouldboard ploughs: Fig.37).

Fig.36

Open-mouldboard plow
(Source: Pellizzi, 1986)

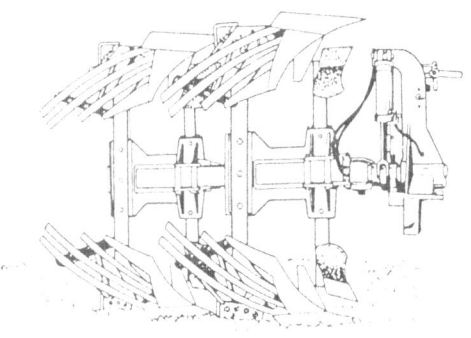

Fig.37

Diamond-type mouldboard plow
(Source: Cavalchini, 1985)

The use of these new tools is spreading gradually, mainly in central Europe (France and Germany) and, in comparison to conventional ploughs, have demonstrated their ability to save a minimum of 12% of energy (in some cases up to 30%).

These tools are already on the market but several research programs are investigating plough lubrication and ploughs made of materials having a low friction coefficient.

The first research line, already in existence at the beginning of the century, investigates the possibility of using air or liquid lubricants. Using air a reduction of 20% of the tractive force has been obtained, but this system requires complex implements which need, in their turn, high energy imputs. More promising is the use of water (Fig.38): energy savings of 10-20% have been obtained using 600 liters/h of water.

<u>Fig.38</u>

<u>Tubing that carries lubricant to the plow bottom</u>
(Source: Schafar and others, 1979)

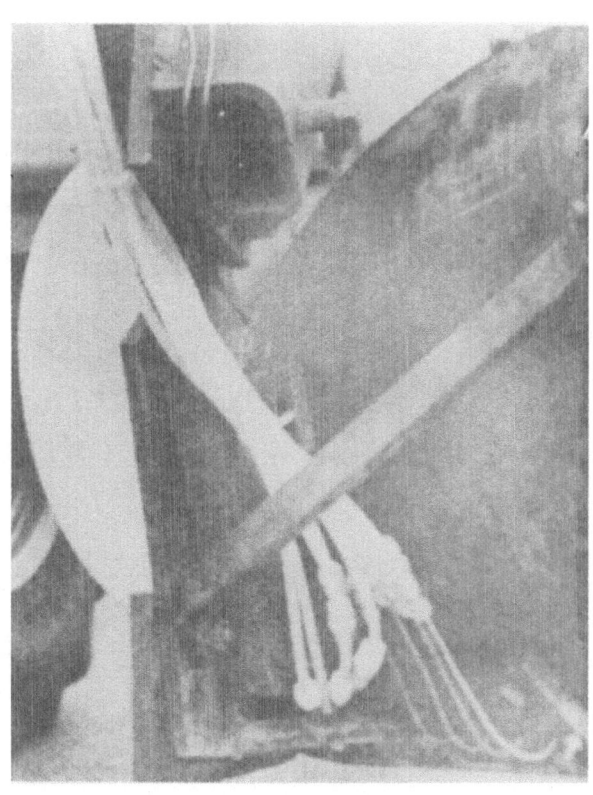

Experiments with low friction materials, by coating the tools with, for example, teflon, have otained a 40% reduction in the tractive force.

Other alternatives use vibrating systems, operating at either low frequency or high frequencies (ultrasounds).

Besides research into the design and construction of tools, many programs have been carried out in order to investigate the possibility of adopting tilling techniques other than ploughing.

In Italy, for instance, in clayey soils having medium consistency, other techniques than deep ploughing have been tried. Two layer techniques working in two passes (subsoiling followed by a light ploughing: Fig.39) or in one single pass (ploughs equipped with a subsoiler: Fig.40) have been perfected that result in energy savings of up to 20-40%.

Fig.39

Different types of subsoilers
The bone B is equipped with two small wings on the bottom
(Cavalchini, 1985)

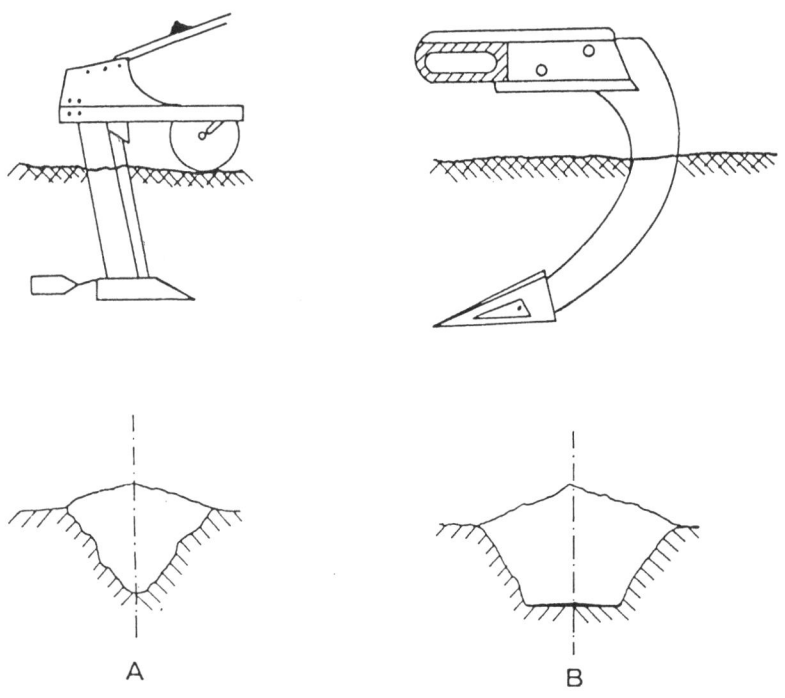

A   B

Fig.40

Plow equipped with a subsoiler
(Source: Scalmana)

Further savings may be obtained through the two layer technique, by utilising open-mouldboard ploughs or diamond-type mouldboard ploughs, but the experiments using these techniques are still at an early stage.

It has to be pointed out that the choice of a suitable tractor - in such a way as to optimize the couple tractor/implements - leads to savings which are not linked with the soil tillage techniques (Fig.41).

In the future, the ploughing energy requirements will decrease mainly because of the reduced tillage depth of cereal crops. For this type of cultivation many investigators -mostly in the northern Europe- agree that a tillage depth close to the sowing depth is sufficient, and thus recommend a reduction of the tillage depth by more than 50%. If we consider that, as a first approximation, the tillage resistence -at the same working width- is proportional to the depth, it is possible to halve the resistence and achieve a more effective coupling of the implements with two wheel drive tractors.

Fig.41

Fuel consumption for different soil tillage techniques using
tractors of different power
(Source: Cavalchini, 1985)

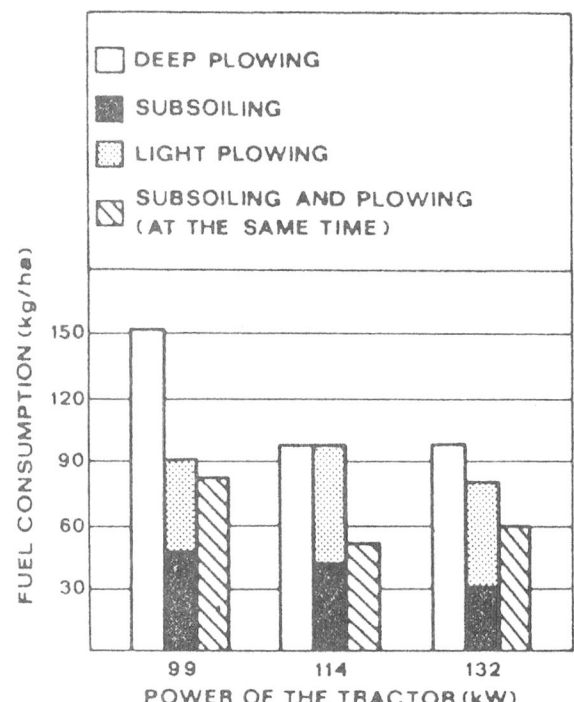

To adopt this technique it is necessary to have a less
compact soil and large cross section tyres. In addition, the
area should be infestations free and have good drainage. If
these conditions do exist, the primary tillage energy spend
could be 60 MJ/ha, that is only 20-30% of the present energy
requirement. To obtain a wide utilization of this technique,
the reduction in the size (and of the weight) of the
different implements must be considered as an important
objective.

Some experiments are in progress at NIAE, in order to
investigate the possibility of using lightweight tractors
equipped with low pressure tyres to prevent damaging the
soil structure and crops. The results obtained so far are
promising in terms of lower energy imputs, but the
investigations are still in their early stages.

Another line of research in Italy (Cavalchini et Al.) is
exploring the possibility of sowing a second crop before

harvesting the main one. This development requires the use of sod-seeding machines and has two objectives: to increase the productivity via an early sowing and to reduce the energy input through different types of tillage, sowing and irrigation. Similar investigations are in progress (Giardini) on sod-seeding and minimum tillage techniques for maize culture.

For root-crops new low energy techniques based on rotary diggers have been examined (Chamen). In <u>table 16</u> data has been gathered concerning sugar-beet cultivation showing a comparison between various techniques.

<u>table 16</u>

<u>Effects of tillage methods on the yield of sugar beets</u>
(Source: Matthews, 1979)

Primary tillage method	System Net rate of work (ha/h)	Mean yield of plots (tonnes/ha)	Degree of 'Fang' formation, % of roots in grade			
			(%)	(%)	(%)	(%)
			1	2	3	4
Mouldboard Plough, 215 mm depth	0.19	31.5*	95	3	2	0
Rotary digger, rotor 100 mm, tines 200 mm depth	0.28	33.5*	85	10	3	1
Rotary digger, rotor 150 mm, tines 250 mm depth	0.27	32.8	86	11	2	1

*Differences not significant
Fang grade 1 No 'fanging' - single tap root
             2 One main root and one subsidiary
             3 One main root and two subsidiaries
             4 One main root and three subsidiaries

In The Netherlands, it has been demonstrated that the reduction and control of the tyre treading due to the machines may lead to an energy saving in the primary tillage stage as high as 50%.

3.2.2 - Rotary hoes and spading machines

For many years rotary hoes (Fig.42) and spading machines (Fig.43) have replaced the plough when the tillage depth is less than 20-30 cm, remaining around 10-15 cm.

On the basis of what we have discussed in § 3.2.1 about the reduction of the primary tillage depth, the use of these types of machines appears promising.

Fig.42

A rotary hoe
(Source: Pellizzi, 1986)

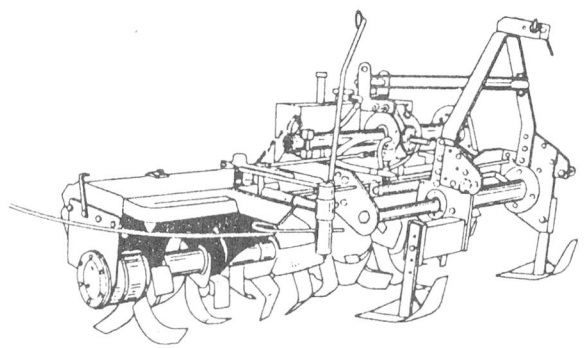

Fig.43

A spading machine
(Source: Pellizzi, 1986)

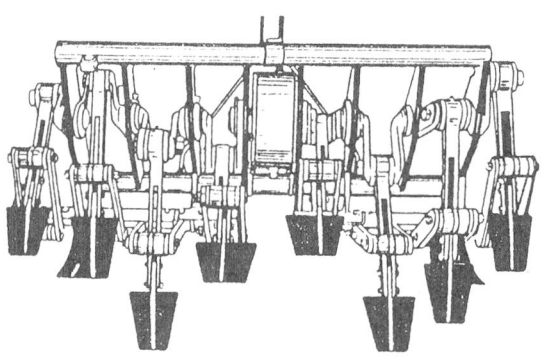

In comparison to the plough, these machines have the following advantages:

- the powering of the operating devices through the power take-off allows the possibility of coupling with light two wheel drive tractors resulting in a more efficient utilisation of the power (15-20% better) in comparison to the use of the traction force;

- the central back coupling of the machines with the tractor makes it possible to keep the tractor wheels on the hard untilled soil and therefore improves the global efficiency;

- better crumbling of the soil in comparison with ploughing: 3-5 times better for the spading machines, up to 30-40 times better for the rotary hoes. The secondary tillage is reduced and, in some cases, eliminated.

However, the turning of the soil and the burying of the soil skin and fertilizer are poorer. This may be armful particularly in clayey soils. The problems of soil compaction must also be considered (Fig.44).

Fig.44

The use of rotary hoes with straight tines prevents the working compaction of the soil
(Source: Pegoraro)

In terms of energy consumption, the rotary hoes require more energy per hectare than ploughs, while for spading machines the energy consumption is equivalent or lower.

Therefore, initially these machines do not appear to have any advantage from the point of view of energy requirements. However, the following factors still apply: better efficiency of the couple tractor/implement (5-10% better vs. plowing), the decreased soil compaction (indirect energy saving) and the better soil crumbling.

The last factor can halve the secondary tillage operations and, consequently, also the corresponding energy cost.

Finally, the use of these machines in place of mouldboard ploughs, may reduce the energy cost of the seed-bed preparation (excluding the traction losses) to 170-180 MJ/ha in clay soils, and to 140-150 MJ/ha in sandy soils. Presently, the seed-bed preparation costs, when using mouldboard ploughs are estimated, respectively, at 300-400 and 180-200 MJ/ha.

3.2.3 - Secondary tillage

For some years the prevailing custom has been to use secondary tilling machines with a high working capacity.

Among the types of machines presently available, a significant development has been achieved for machines with rotary tools powered by the power take-off (rotary diggers, rotary harrows: Fig.45).

More recently, particularly in the Mediterranean area, other types of implements are being used which possess a combined set of tools (Fig.46) and are able to carry out the complete seed-bed preparation in one single pass.

The combined machines have proved, in comparison with machines powered by the power take-off, to have a better performance and be more energy efficient. The energy savings are around 15-20%, but they can be as high as 50% if the energy spend is calculated in terms of the volume of ground handled.

83

Fig.45

A rotary harrow
(Source: Cavalchini, 1986)

Fig.46

Multitiller used to obtain a complete seed-bed preparation
in a single pass
(Source: Cavalchini, 1981)

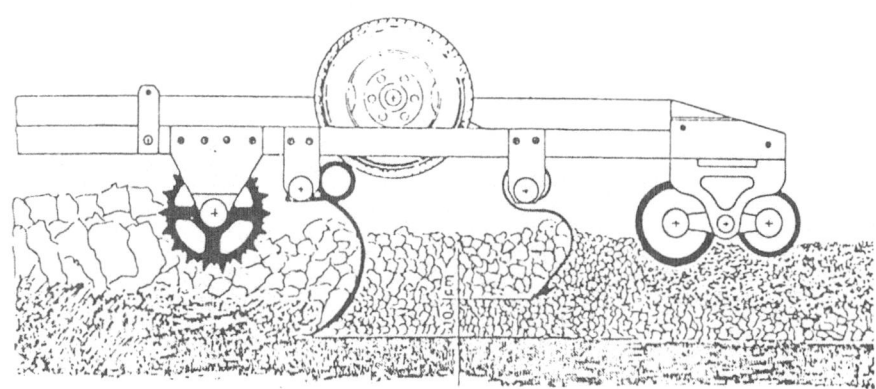

3.2.4 - Operative conclusions

The analytic survey we have carried out indicates that the soil tillage techniques are ready to adopt many of the innovations previously described. The non-necessity of deep tillage has been proved and the advantages of shallow tillage have been discussed. Lighter tillages imply fewer field passes and therefore less fuel consumption.

In the medium term, a significant uptake and utilisation of the implements and techniques already outlined is expected.

In the coming ten years, it is possible to anticipate a global reduction of fuel consumption by around 30% corresponding, at the EEC level, to 1.5 Mtoe/year. It must be recognised, however, that this result can only be obtained in this period if incentives and information are given to the farmers, particularly in the mediterranean regions and in the internal areas.

Further energy savings could be achieved, in the long term, through the development of techniques suitable to control the movements of vehicles in the fields. A higher technical preparation of the operators and an upgrading of the farm structures are necessary. Therefore, the spreading of these processes will require time and a strong commitment.

3.3 - Auxiliary machines for crop cultivation

Farming operations other than soil tillage and crops harvesting (fertilizing, sowing, transplantation, treatments etc.) account for about 20% of the total direct energy requirement of the crops. The multi-step nature of these operations makes the energy spend recorded for each single operation very small, so that the fuel savings attainable by the development of new techniques does not appear as significant.

However the savings, in terms of indirect energies supplied to the various cultures are significant, particularly in the distribution of chemical fertilizers, pesticides, weedkillers and seeds.

Even though this aspect of energy consumption is not included in this survey, it seems worth-while to discuss it briefly and point out the opportunities and the need to carry out a specific survey on this subject.

Savings can be achieved through three different types of action:
- 1st) At the level of the farming operations, by means of suitable technical assistance to the farmers: appropriate soil analysis, careful specifications on the real fertilization needs of the various cultivars, definition of clear "calendars" for the different operations;

- 2nd) At the level of the manufacturing of the machines, by means of careful experimentation and production of some new technologies that are able to induce considerable savings in the distribution of several chemical and biological inputs, without negative effect on the cultural yields, and in some cases able to replace completely the inputs themselves;

- 3rd) By increasing the research on integrated biological disease control and on the nitrogen fixation process.

In the following pages we will deal only with the improvements made possible by the introduction of new mechanical technologies.

3.3.1 - Machines for fertilizers distribution

Possible energy savings can be achieved through the use of:

- machines allowing a localized distribution of pelletized

fertilizers (Fig.47) with savings of fertilizer between 20
and 50 % versus the generalized scattered broadcast,
depending upon the type of crop, its plantation settling and
the type of ground;

<u>Fig.47</u>

<u>Pneumatic seed drill equipped with fertilizer distributor
and microgranulates applicator to the rows</u>
(Source: Gaspardo)

- pneumatic spreading machines (Fig.48), in comparison with
the present systems based on the centrifugal reaction in the
spreading discs, can save around 10-15%. In the spreaders
presently used it would be worthwhile to introduce systems
to control the flow and the working width, so that the
driver can improve the utilization of his machine.

These short discussions indicate the importance of this
subject and that it would be worth-while to consider
demostration and promotional actions.

Fig.48

Pneumatic fertilizer spreader
(Source: Kuhn)

3.3.2 - Pesticides and herbicides spreaders - Alternative solutions

In this sector we discuss the different possibilities offered - in liquid applications- by new developments:

- electrostatic distribution, where the liquid is fed by gravity (Fig.49). A ring-shaped electrode electrically charges (positive pole) the liquid, producing atomised droplets of an uniform diameter. The droplets adhere to the crop leaves through elecrostatic attraction.

88

Fig.49

A) Electrostatic sprayer. B) Scheme of the system.
(Source: Pellizzi, 1986)

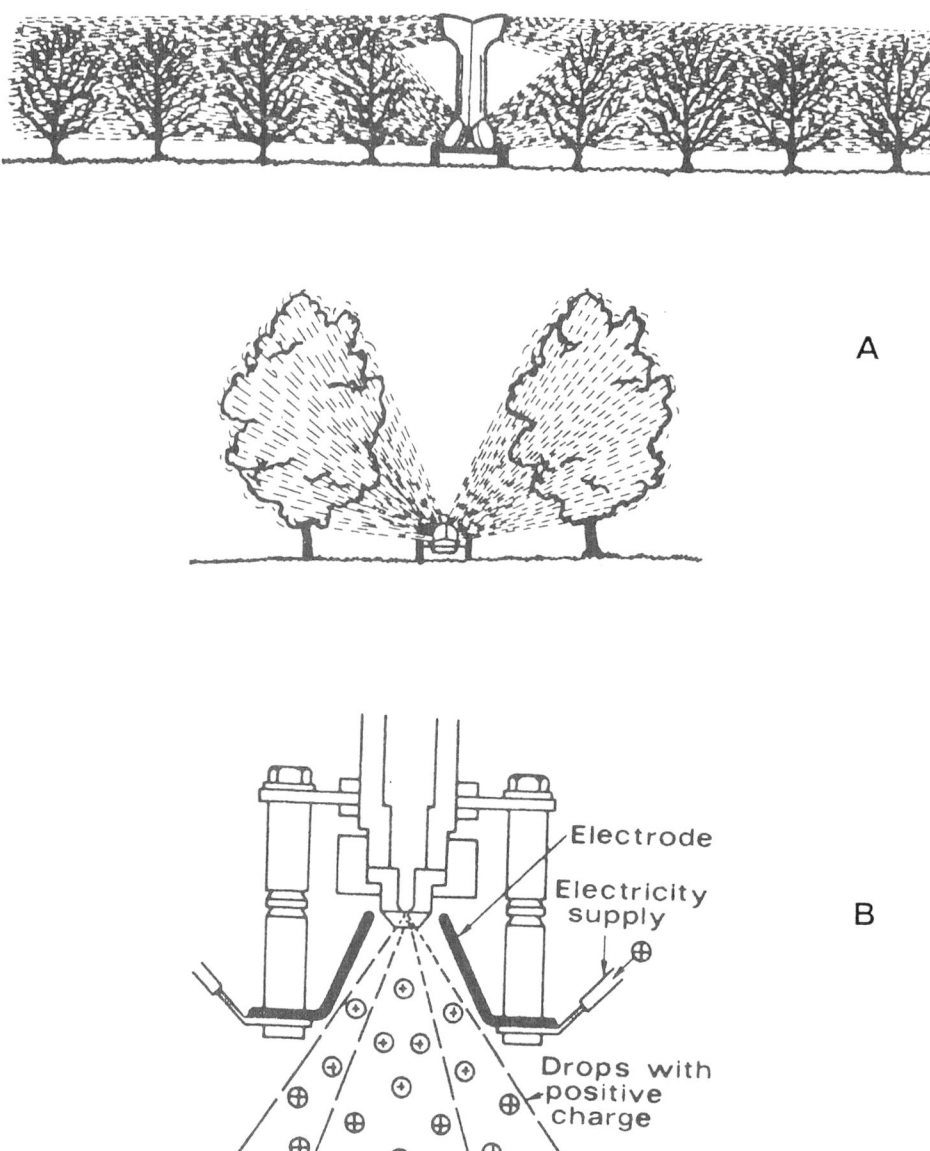

A

B

Electrode

Electricity
supply

Drops with
positive
charge

89

- Contact distribution (Fig.50), based on a distribution
device formed by ropes or clothes drenched by capillary
action with the active liquid delivered by a special tank.
When the machine comes forward, it touches and moistens the
tallest weeds.

Fig.50

Herbicide sprayer of contact type
(Source: Pellizzi, 1986)

Distribution with a recycling system (Fig.51): the
nozzles spray horizontally onto the weeds but the residual
liquid falls on a recovering shield from where it is
filtered and recycled.

Following the experiments carried out on these systems
in different European countries, this type of system could
provide a saving in the active products required of around
20-30%.

In the field of the pesticide distribution, there are
also many control devices which are based on the speed of
the machine and the flow and pressure of the liquid
(Fig.52). The simplest systems provide the operator with an
indication of the quantities used per unit of treated
surface, while the more sophisticated systems automatically
control the flow and the pressure in such a way that the
quantity of active product spread per unit of treated
surface remains constant.

Fig.51

Sprayer with herbicide recovery system: 1, tank; 2, pump; 3, powered valve; 4, nozzles; 5, recovering device; 6, collectors; 7, filters; 8, recovery pump.
(Source: Pellizzi, 1986)

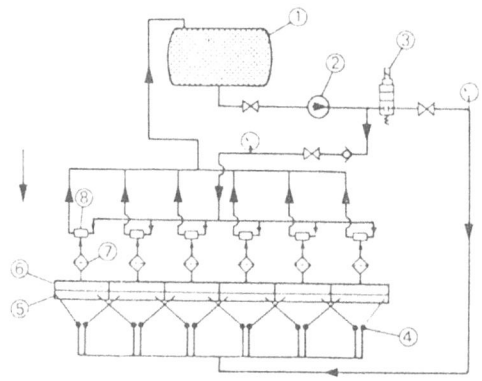

Fig.52

Electronic control system for a sprayer
(Source: Dickey John Co.)

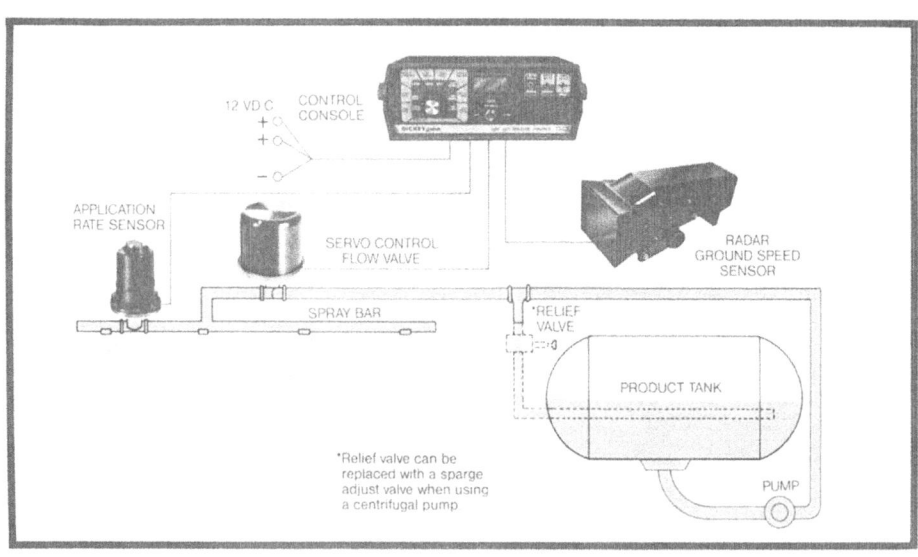

A completely different system, still in the development phase, could completely replace the use of chemicals. It uses electromagnetic microwaves produced by a high voltage (8-10.000 V) generator. These microwaves generate a large quantity of heat and, if concentrated on the soil and the cultures, cause the parasites cells to burst or the pests to be inactivated and sterilized. The microwaves have an effect in the soil to a depth of 7-8 cm., but the ground must be dry and at a temperature higher than 10°C to be effective.

For the distribution of pesticides on herbaceous or arboreal cultures, the use of pneumatic sprayers must be recommended. This type of sprayer works at a very low water volume (less than 30 l/ha), saving 10-20% in comparison to conventional systems. These figures can be further improved by using the electrostatic distribution device.

Finally, we must discuss the possibilities offered by the use of motor-gliders (30-40 kW) equipped with tanks (80-100 l) which can be used instead of the ground operating machines. This solution (<u>Fig.53</u>) considerably cuts the working time, with direct (8-10%) and indirect (12-15%) energy savings.

<u>Fig.53</u>

<u>Glider equipped with a spraying machine, propelled by a 30-40 kW engine</u>
(Source: De Zanche, 1985)

3.3.3 - Transplanting and seeding machines

Today, transplanting is completely neglected in Europe, but this sector was once typical in rice growing and in horticultural crops.

Nevertheless, recent experimental investigations have shown that mechanical planting of small plants coated with clods of earth has some advantages:

-to increase the yield,
-to decrease pesticide distribution,
-to reduce the soil engagement and enable two crops per year to be harvested,
-to require less water for irrigation.

As an example, the use of this technique in rice growing (500,000 ha in EEC) has the following advantages (in comparison to direct sowing):

-reduction of the ground occupancy by about 30 days;
-reduction of the water consumption by 20%;
-reduction of pesticides and weed-killing treatments by 30-40%;
-yield increase by 6-15%.

These advantages are not fully realised because there are no reliable machines on the market having a high working speed combined with low manpower requirement.

The existing transplanting machines are of Japanese origin and, even if their productivity is not satisfactory, the data we have presented should encourage demonstration actions, which may motivate the farmers and, at the same time, promote the development and the optimization of more suitable machines.

There are some innovative techniques for wheat (and for the other winter-autumn cereals) that aim to utilise some of the principles already adopted for maize and sugar-beets.

Demonstration actions are necessary to promote the use, in these crops, of sod-seeding machines (Fig.54) and minimum tillage techniques, testing these innovations in different soils, different climates, and on various vegetable species.

Fig.54

Sod-seeding machine
(Source: Gaspardo)

3.3.4 - Operative conclusions

As we have already mentioned, the possible energy savings in cultivation operations are few and difficult to evaluate.

However, in this field, considerable indirect energy savings are possible, through a more effective use of existing products and means. This is the reason why we recommend a specific multi-disciplinary survey on this subject, also taking into account the environmental aspects of the use of agro-chemicals.

3.4 - Harvesting machines

As mentioned in § 2.11, the harvesting operations use 32% of the total energy requirement, and represent the second largest fraction of the energy spend.

In contrast to what we have seen in § 3.2 about soil tillages, the possible energy saving opportunities are reduced, because the harvesting systems are numerous and variable so that a global evaluation becomes impossible.

3.4.1 - Grains harvesters

Until recently all combine harvesters had the same operational scheme, but today one must distinguish between conventional and axial machines.

According to Kutzbach, the axial type machines require more energy because of the straw crumbling caused by the axial threshers.

A survey on about 100 machines working on different products did not indicate any significant differences. At present, we consider it would be difficult to obtain energy savings through the use of alternative threshing systems.

Theoretically, it would be possible to save energy by replacing the hydrostatic transmissions presently used by means of mechanical drives. But it must be pointed out that the power used for the self-displacement of the machine represents only 20% of the total and the advantages of the continuous speed variation offered by the hydrostatic drive are too important and cannot be given up.

Energy savings could be obtained by using monitoring systems and automatic electronic controls. On this aspect a large bibliography exists and many systems are already in use.

One of them is the monitoring of grain losses. This device is based on piezo-electric sensors mounted on the outlets of the separation organs and it is able to count how many grains fall to the ground. In the simplest systems the signals enter a central decoder which indicates, on a suitable display, the general trend of the losses. In more sophisticated systems, the speed of the machine is also recorded and the loss monitoring is referenced to the harvested surface (Fig.55).

Fig.55

Combined grain loss monitoring system
(Source: Dickey John Co.)

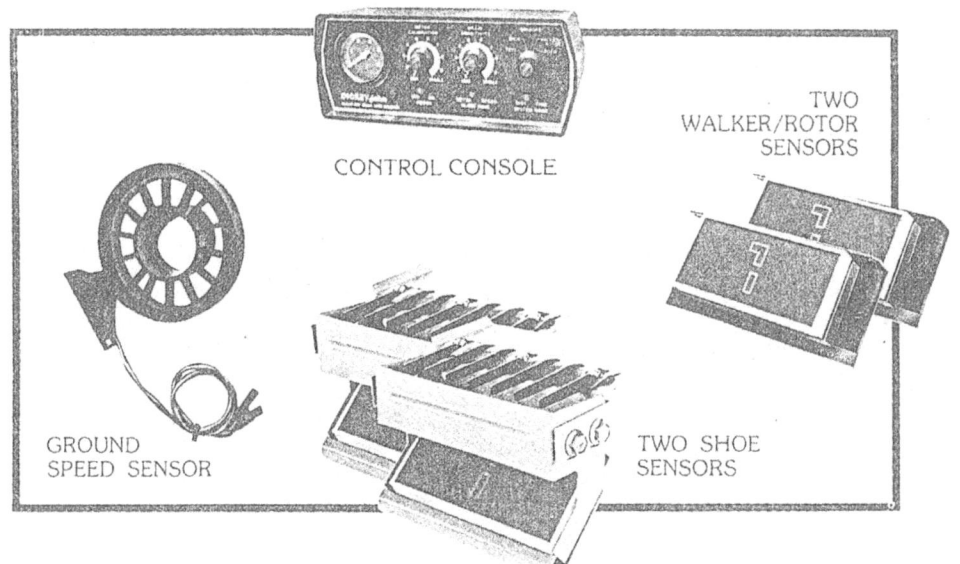

CONTROL CONSOLE

TWO
WALKER/ROTOR
SENSORS

GROUND
SPEED SENSOR

TWO SHOE
SENSORS

This type of device makes it possible to choose the most
suitable operating conditions and by controlling the advance
speed optimizes the performance of the machine.

The next developments will utilize the continuous
variation in advance speed and thresher rpm, to produce a
full automatic control system incorporating these two
functions.

Fekete and Földesi have shown that under Hungarian
conditions these systems can increase the productivity by
15% and reduce the specific consumption by 10%.

According Huisman, these automatic control systems are
too expensive if we take into account that combines are
already efficient, particularly in the high productivity
lands of continental Europe.

Claas Co. markets combines and choppers for maize
harvesting with an automatic steering control based on the
use of two groping sensors.

A Japanese company (Iseky), has proposed a model of a
small harvester-thresher that is able to follow the edge of

the culture, once the field has been opened. A special device, controlled by a computer, operates the steering and, according to the manufacturer, no on board operator is required.

Finally, over the last five years working methods based on the harvesting of the whole plant have been proposed, leaving the separation of the grains from the straw to a second fixed point operation. Initially this system was proposed by a Swedish company who built special harvesting machines similar to the field-choppers, and a separation-drying plant using the harvested straw as fuel (Fig.56).

Fig.56

Scheme of a whole crop harvesting system

1) Harvesting using a self-propelled chopper (180 kW) equipped with quick on-off mounting containers; 2) handling through 40 m3 containers; 3) drying and threshing at stationary fixed point; 4) straw processing: light fraction can be used as animal fodder, other parts for energy or paper.

(Source: Kockums, 1977)

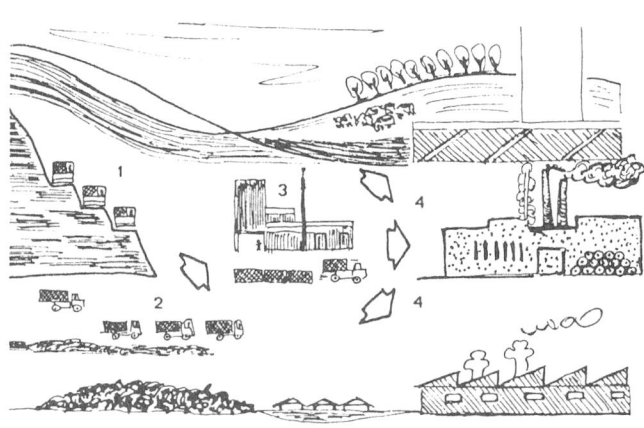

In the north European countries the harvesting time is very short, occurring at the end of the summer, so that the crops have to be dried: in these conditions the system seems profitable in terms of energy requirements.

However, at the moment, a total reorganisation of the traditional working scheme would be too expensive and not balanced by the advantages of the system. In addition, tests and evaluations carried out in France and Italy, did not confirm the advantages of the method. In Italy, where a traditional scheme has been compared with two schemes of integral harvesting, the energy spend per unit of harvested grains appeared lower when the combines were used (table 17).

table 17

Comparative results obtained with different mechanization chains for cereal harvesting

Working chaine	Power avai-lable (kW)	Power used (kW)	Mass (t)	Work pro-duc-tivity (man.h) ha	Grain har-vested (t)	Work pro-duc-tivity (man.h) t	Straw	Total prod. harvested (t)	Total prod. harvested (man.h) t	Energy cost grain (kWh/t)	Energy cost prod. (kWh/t)
Combine	285 (100)	165 (100)	12.3 (100)	3.28 (100)	4.35 (100)	0.66 (100)	2.76 (100)	7.71 (100)	0.42 (100)	80 (100)	51 (100)
Field chopper	170 (100)	118.5 (71.8)	15.2 (124)	4.00 (134)	4.63 (93.5)	0.86 (121)	4.14 (150)	8.77 (114)	0.46 (107)	98 (122)	51 (100)
Cutting and self-loading trailer	139 (48)	83 (50)	18.2 (148)	5.34 (179)	5.05 (102)	1.05 (160)	4.50 (163)	9.55 (124)	0.52 (131)	97 (121)	52 (102)

(Source: Cavalchini et Al., 1984)

3.4.2 - Forage harvesters

The harvesting techniques used for meadowland forage harvesting are:

-traditional haymaking,
-two stages haymaking,
-dehydration,
-ensilage.

The working chains used are very different, as are the energy input-output characteristics for each technique.

Control of the nutritional value loss is essential, and good results are obtained by a proper choice of the machines and of the working schemes. It appears from table 18, that the highest efficiencies are obtained with ensilage and traditional haymaking. The two step haymaking process can give an equivalent efficiency if solar energy is utilized.

table 18

Energy efficiency of the different techniques for harvesting alfalfa and maize
(Source: Cavalchini 1977)

Forage	Harvesting method	Energy cost (GJ/ha yr) (1)	Energy produced (GH/ha yr) (2)	Process efficiency (2/1)
Alfaalfa	Traditional hay making	9.04	24.5-44.5	2.82-4.92
	2 times hay making	28.00	38.9-55.6	1.39-1.98
	Silage	10.39	31.8-52.5	3.06-5.05
	Partial dehydration	109.34	46.1-58.5	0.42-0.53
	Total dehydration	180.34	56.6-65.2	0.31-0.36
Maize	Silage	30.2	70-80	2.32-2.65

Dehydration, either partial or total, is costly in terms of energy spend, and total dehydration presently does not produce any economic advantages.

We want to point out that some operations (mowing,

haymaking, swath-making) require a very small power input but, in practice, the power is often supplied by engines with large power capacities. The fuel consumptions are, consequently, higher than for a suitable coupled engine-operating machine.

Among the balers, the traditional high density pick-up balers require 1-1.2 kWh/t, the roto-balers 1.5 kWh/t, the big-balers preparing rectangular high density bales 1.5-1.7 kWh/t. Therefore, in terms of the energy requirement, both the big-balers and the roto-balers are not to be recommended. Nevertheless, both round balers and big balers have advantages from the standpoint of productivity, time saving and reduction of labour.

The large balers (4-5 t) need high power and heavy handling machines (100 kW) which tend to increase soil compaction problems. However, this type of baling permits significant savings if the transportation range is great (<u>Fig. 57</u>).

<u>Fig.57</u>

<u>High density big baler: it allows a better use of the handling equipments</u>.

When ensilage is practised, choppers (both self-propelled and tractor mounted) equipped with headers are necessary to obtain good fermentation and preservation of the product. The energy consumption of these machines is around 2-2.2 kWh/t, i.e. much higher than for conventional balers, if we take into account that the product moisture content is higher. Nevertheless, this technique, as with the two step haymaking process, is essential in the northern Europe and also appears to be technically more efficient in central and southern Europe.

There are few opportunities of direct energy savings for maize silage. In fact, only by chopping in longer shreds would it be possible to save energy (Fig.58). But, technically, only cutting length shorter than 6-8 mm can guarantee a good quality silage.

Fig.58

Forage harvesting power absorption with different cutting lenghts
(Source: UMA, 1976)

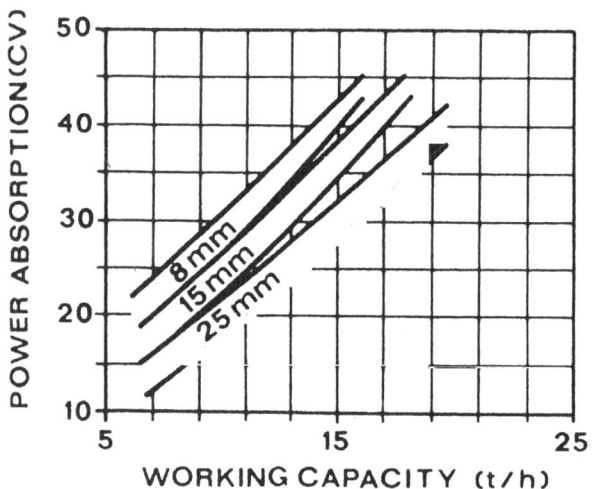

Indirect savings are possible if the soil compaction is reduced. Therefore, multiple-axles lorries with large tyres must be recommended. The "Field Queen" model made by the Hesston Co was equipped with large tyres and large cases. This machine can move on fields that are unsuitable for conventional vehicles.

3.4.3 - Sugar-beets harvesters

Some differences in energy consumption are evident when comparing the two working schemes (split or combined operations) typical of beet harvesting.

With split operations it is possible to save about 15% (Baraldi) or up to 40%,according our records. Further indirect savings are associated with the split operation scheme, i.e. reduced soil treading due to the larger working width. In the split method the diggers cover 6 or more rows, whilst in the combined working scheme only 1 or 2 rows are harvested in each pass.

Today trend is towards the split work scheme in which the working capacity and the time saving are higher. But there are few opportunities for energy savings in this scheme.

However, significant indirect energy savings can be achieved if the leaves are also collected as fodder in the same operation. In this case, the energetic efficiency of the culture is increased by 15-20% (table 19).

table 19

Process efficiency in harvesting sugar beets
(Source: Cavalchini et Al., 1984)

| Harvested products | Harvesting methods | Energy cost | Energy produced | | | | Process effincency |
			roots (GJ/ha)	byproduts for animal fodder (GJ/ha)	byproduts for fertilization (GJ/ha)	total (GJ/ha)	output/input
Roots	2 passages	19.2	109.2	-	9.6	118.8	6.06
Roots	1 passage	21.0	109.2	-	9.6	118.8	5.66
Roots+ byproduts	2 passages with chopper	21.4	109.2	31.4	7.6	148.2	6.92

3.4.4 - Potatoes harvesters

For sugar beets and potatoes the harvesting phase energy spend can be reduced if a split working scheme is used and machines that are able to dig several rows, instead a single one, are employed.

Unless the harvesting operations are jobbed, the choice of operation is dependant on the size and organization of the farm.

According to our evaluations, the two-rows diggers working with the split scheme allow energy savings of about 40-50% in comparison to the consumption of the single-row semi-automatic machines (130-150 MJ/ha versus 260-300 MJ/ha).

This energy saving will be less than stated above because of the energy imputs necessary for the fixed point sorting, even though these energy spend are small. We point out that this working scheme guarantees a better quality final product.

3.4.5 - Fruit harvesters

Today, fruit harvesting is by hand, though some areas utilize the same types of auxiliary carts which are also used for pruning and thinning. Integrated mechanical harvesting is not yet feasible on a large scale.

Even though these machines do not carry out any real harvesting operation, only allowing the operators to approach the working area, their fuel consumption is high and represents 10-15% of the total. This is due to the long time required for the fruit picking and also because of the long waiting times when the machines are idle but the engines are still running.

For harvesting carts we can assume that significant energy savings are possible if the present systems are replaced by battery powered equipment.

Besides the energy savings, this solution presents many ergonomic advantges, for example, the lack of noise, of vibrations and of the chemical pollution due to the exhaust gas.

Tests in Italy by Petrone have indicated the potential of an electric powerlift for citrus harvesting, where savings up to 60-80% have been recorded in comparison to conventional systems (Fig.59).

103

Fig.59

Block diagram of an electric powerlift for citrus harvester:
ME1, ME2, DC motors; MRE, DC motor with brake and reducing
gear box; A, accumulator; D, differential; P, hydropump;
S,others
(Source: Petrone, 1981)

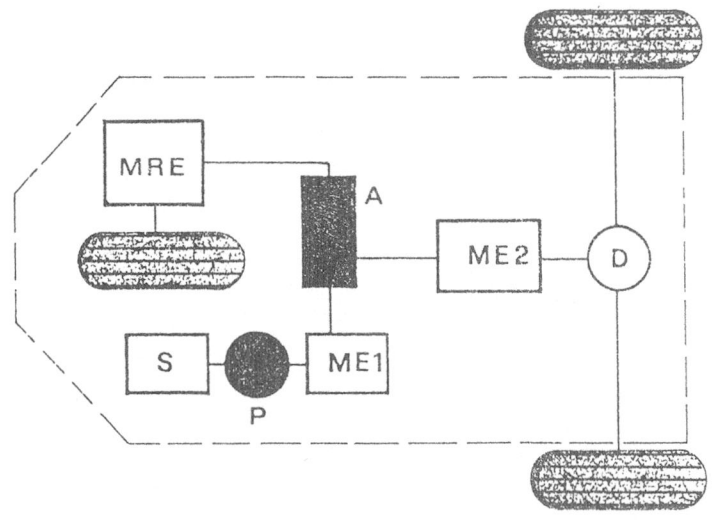

A new research program (CESMA and Istituto di Ingegneria
Agraria of the Milan University) in Italy is now beginning
to transfer the electric system to the auxiliary carts used
in most of the arboreal cultures.

For several years the CEMAGREF has undertaken a reserch
program concerned with the robotisation of fruit harvesting
based on chromatic vision devices. It is too early to
determine if there is a future in developments of this type,
mainly because very sophisticated equipment are used on very
cheap products (the single fruits) and it is likely that
the costs will be high and the working efficiency very
limited.

3.4.6 - Grapes harvesters

In France, mechanical grapes-harvesting is very widespread (6.500 machines) whilst in the other EEC countries mechanization is just beginning. In Italy, for instance, where the viticultural area is practically the same as in France, only 300 vine-harvesters are operating.

As we pointed out in § 2.6.2, there is practically no difference, in terms of energy consumption, between the mechanical and manual harvesting. This is because the manual harvesting is time consuming and, in the vineyards, the internal movements require the use of the operating machines for long periods.

In the future, mechanical harvesting, in Italy and also in the other EEC countries, will be based on the use of the 3rd generation harvesters, recently available on the French market. This type of harvester is basically a bogie frame (tractor coupled or self-propelled) on which several types of equipments (for harvesting, for treatments, for tillage, for pruning, etc.) can be mounted (Fig.60).

Fig.60

3rd generation vine-harvester
A self-propelled frame on which it is possible to mount different types of implements
(Source: Braud)

Electronic control equipments has also been installed on vine-harvesters, for example, the automatic control of the advancing direction. This type of equipment should lead to a maximum working efficiency and reduced consumption.

At present, it is difficult to evaluate the future impact of these systems on the operative conditions. Because of the higher power required and the different utilization diagram (continuous and almost steady), the electric systems proposed for fruit harvesting are not suitable for vineyards.

In our opinion, considering the variety of the functions and the complexity of the transmissions (presently mostly of the hydraulic type), it would be interesting to investigate the potential of hybrid diesel + electric power systems. This type of system could lead to fuel savings of around 20%.

3.4.7 - Olive harvesters

Olive-harvesting is mostly by hand, with the aid of collecting nets.

The mechanical harvesting is based on the use of a shaker. This implement can take many forms, following different principles, and the energy requirements may also be different. Because of the variety of models, it is difficult to carry out evaluations on the energy utilization.

When the shaker is applied to the tree for only a few seconds, the efficiency of the powering engines is low.

There is an Italian shaker (Consiglio Nazionale delle Ricerche) (Fig.61) which is powered by the power take-off of a tractor via an oil and air accumulator which is continually charged, even when the tractor switches from one tree to another.

This principle allows the use of tractors having 3-4 times less power than a conventional shaker while still possessing the same shaking power. Table 20 indicates the reduced energy requirements.

This solution could also be used for harvesting fruit for industrial processing.

<u>Fig.61</u>

<u>Schematic view of a shaker equipped with an innovative
energy storage equipment</u>
<u>(Source: Baldini, 1979)</u>

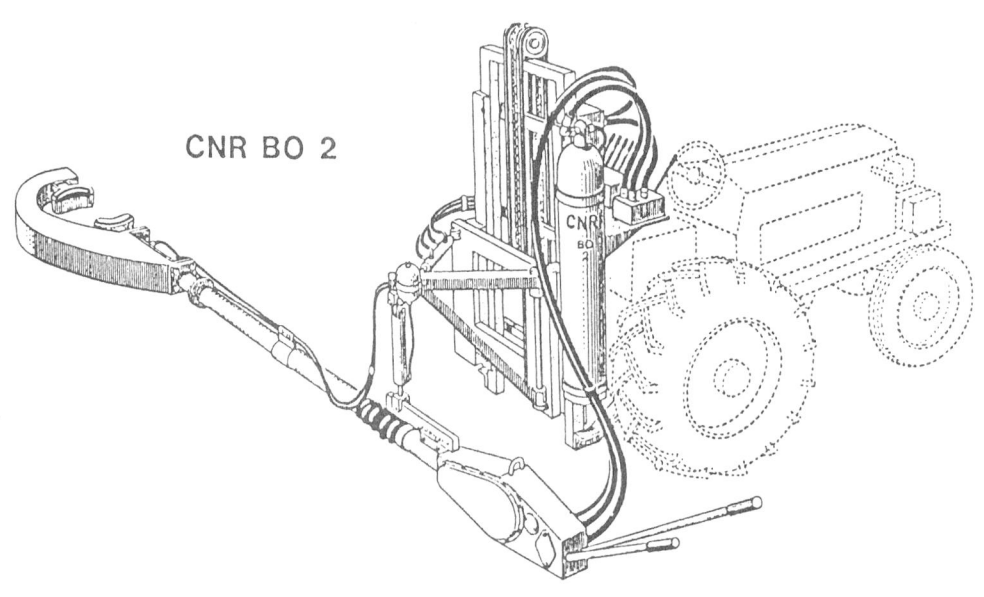

CNR BO 2

<u>table 20</u>

<u>Comparative fuel consumption between a conventional
self-propelled shaker (A) and an innovative shaker equipped
with energy storage system (B)</u>

Type of machine	Power (kW)	Transfer time (s/tree)	Shaking time (s/tree)	Fuel consumption (g/tree)	Index number
A	97	60	30	213	100
B	55	60	30	180	84

3.4.8 - Operative conclusions

The harvesting machines sector is highly diversified and the machines and implements used are similarly very different.

In summary, for cereals and other grains crops it is possible to obtain savings of around 10% through the use of monitoring and automatic control devices on the combines. Besides the improved efficiency of the machines, the working conditions of the operators have been improved.

In forage harvesting a slight increase in the global consumption will probably be recorded, because traditional haymaking is progressively being replaced by alternative two step haymaking techniques and by ensilage. These techniques have higher specific energy demands, but lead to an improvement in the quantity and in the quality of the crops harvested. The simultaneous improvement of the machines and the professional training of the operators will lead, in the short term, to specific consumptions equivalent to present levels.

For sugar beet harvesting, an appropriate choice of equipment and a better use of the harvesting machines should induce energy savings of around 8-10% of the present requirements.

Higher savings will be possible in potatoes harvesting where the same type of actions may lead to global savings of around 15-20%.

Finally, in the harvesting of fruits, citrus, grapes and olives, even though for some machines improvements are possible, fuel savings are not predictable. For these cultures, a significant backlog exists in the mechanisation of harvesting operations and it is hard to predict the speed with which the newest systems will be introduced and their impact on fuel saving.

In the harvesting sector, in the medium term, energy savings equivalent to 0.10-0.15 Mtoe/year are foreseen. This figure is much lower than the corresponding savings achieved through the improvement of tractor performances and the development of soil tillage techniques.

4 - TRACTORS AND AGRICULTURAL MACHINES MANAGEMENT:
EVOLUTION AND ENERGY SAVING PERSPECTIVES

Another aspect of the possible direct energy savings in
farming mechanization is related to the proper choice and
management of the machines, to the adaptation of the fields
to the performances of the machines and, above all, to their
working capacity.

The working capacity can be measured in ha/h, once the
width of the working front and the advancement speed are
known.

It is well known that a rational mechanization must be
based on maximizing the utilization coefficients of the
machines performances (either for tractors or for
implements). The considered performances are basically:

- for all the agricultural machines, the useful
lifetime (Vu), that is the number of working hours the
machine can execute before it is wrecked, using normal
maintenance and repair practices. The useful lifetime
must be shorter than the technical obsolescence time
(Ot), that is the number of utilization years before the
machine becomes technically out-of-date and its
replacement is economically profitable;

- for the tractors and the derived machines, the
available power (Pd);

- for the implements, the working capacity (C1).

Useful lifetime and technical obsolescence vary for the
different groups of machines as shown in table 21.
Nevertheless, for each machine it is possible to define the
useful annual lifetime (Va), expressed by the ratio

$$Va = Vu/Ot \quad (h/year)$$

There are economic and energy advantages if machines are
used for a number of hours as close as possible to Va (and
even more).

The highest exploitation of Va, Pd and C1 depends upon a
series of external conditions. These conditions have either
an agronomic, structural, or technico-operative character
and some of them can be modified, at least partially.

table 21

Technological maturity and technical life (Ot)
for different machines
(Source: Pellizzi, 1985)

Type of machines	Technological maturity	O_t (years)
Primary tillage	medium-high	10-12
Seed bed preparation	medium	8-10
Fertilization	medium	8-10
Drilling and transplanting	medium-low	6-8
Pesticide distribution	low	4-5
Irrigation	medium-low	6-8
Forage harvesting	medium-low	6-8
Cereals harvesting	medium	8-10
Industrial crops harvesting	medium-low	6-8
Vegetable harvesting	low	4-5
Fruit harvesting	low	4-5
Handling equipment	medium-high	10-12
Products conservation and conditioning	medium-low	6-8
Fodder preparation and distribution	medium-low	6-8
Milking	medium-low	6-8
Stable cleaning equipments	medium-low	6-8

From the agronomic standpoint the value of Va depends:
on the adopted production regulations; on the biological and
cultural cycles of the different species; on the time
requirements of the various operations in respect of the
different species and cultivars; on the weather conditions
and the pedologic characteristics.

C1 is influenced by the conditions of the crops
(flattened crops; presence of weeds; growing systems).

Several structural factors condition the exploitation of
both Va and C1. We can summarize them as follows:

- size of the farms and, more particularly, of the surfaces
allotted to the different crops. The areas allotted to crops
are often too small for the full utilization of the
different machines.

- size and shape of the fields: as fields get smaller, the

negative effect on energy requirements and working capacity increases, due to the increased incidence of the dead time on the total working time.

Technical-operative conditions have a large influence on the optimization of the above mentioned parameters through:

1)-the rationality of the coupling between tactors and implements, that is the coherence between the power available at the outlet of the powering machines and the power the implements can utilize;

2)-the skill of the driver and his knowledge about the machines limitations;

3)-the existence, on the tractors, of automatic control devices and their degree of sophisification.

A summary is given below:

- for tractors, $\eta u = H/Va$ the hourly utilization coefficient (H is the actual number of yearly working hours of the tractor) and $\eta p = Pi/Pd$ (Pi = average power actually employed and Pd = available power) the power utilization coefficient. The aim is to maximize both coefficients and, therefore, their product ηt:

$$\eta t = \eta p \times \eta u \rightarrow 1.$$

- for implements, if ηu is their hourly utilization coefficient, and $\eta c = Cr/Cl$ is the utilization coefficient of their working capacity (Cr is the actual working capacity), the aim is again to maximize both the coefficients and, therefore, their product nmo:

$$\eta mo = \eta u \times \eta c \rightarrow 1.$$

4.1 - Present operating conditions of the machines

We consider it would be useful to discuss in detail three particular points, that have an important effect on energy consumption:

-utilization of the tractor power (ηp);
-coupling tractors/implements;
-utilization of the working capacity of the implements(ηc).

For the first two points, we must emphasize that, in practice, it is not possible to achieve a total exploitation of the installed power, i.e. $\eta p = 1$. The different operations the tractor must carry out result in different power absortions of the implements coupled to the tractor and some reserve of power (10-15%) must be available to overcome the peak-loads. In the course of the tractors useful life, a progressive deterioration takes place,e specially in the engine. This deterioration can reduce the initial nominal power by 10-15%. Finally, we get

$$\eta p \ max \ = \ 0.7$$

This is the optimum theoretical value. In practice, the average figure for European agriculture is $\eta p = 0.50$. This is due (because of the narrowness of the existing structures) to the utilization of the tractor for numerous scattered operations with different energy requirements and different power absortions.

Few published data are available on this subject at the EEC level, even though the analysis carried out in Belgium and by the NIAE are interesting. In the Italian conditions, we have noticed that a generic tractor is employed with a consumption of 20 kW during 60% of its useful life, during 20% of its working time the required power is 30 kW, only in the remaining 20% of its working time is higher power required. Because the average installed power is around 45-50 kW it is easy to calculate, using the diagrams of Fig.62, the non-necessary consumption.

Through a more suitable choice of engines and of the implements coupled with them or by making use of jobbing utility companies, it should be possible to improve the average value of ηp and, therefore, save 25-30 g/kWh of fuel. Considering that, in Italy, a tractor works, on average, 300-350 h/y, it is possible to evaluate the possible energy savings. They are around 360.000 toe/y and this figure represents 12-13% of the energy consumption of the sector.

The same situation probably exists in the other 11 countries of the EEC and the corresponding energy savings can be evaluated at around 8-10% of the present requirements, i.e., for the total EEC, approximately 1-1.2 Mtoe/y.

Fig.62

Engine specific consumption at different loads
(Source: Dufey, 1984)

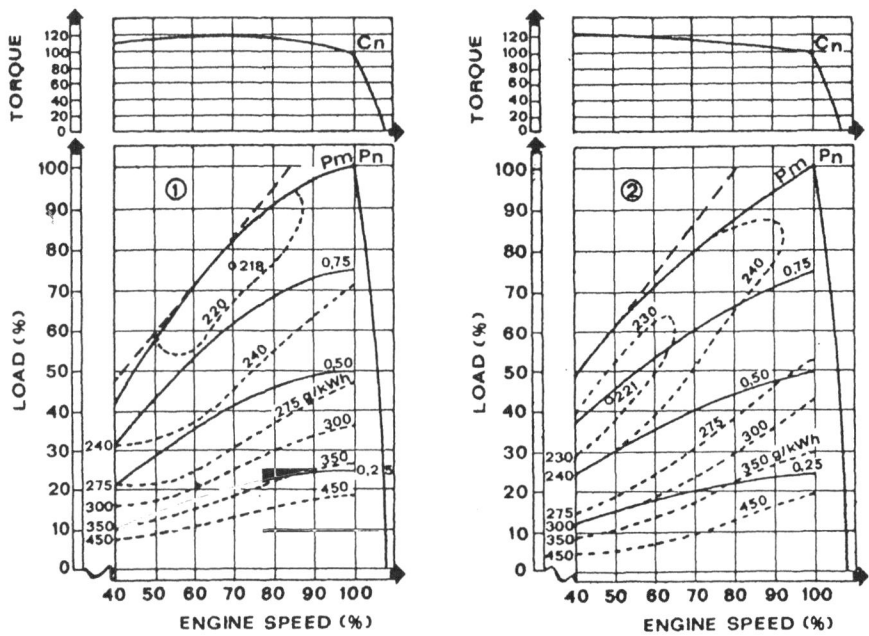

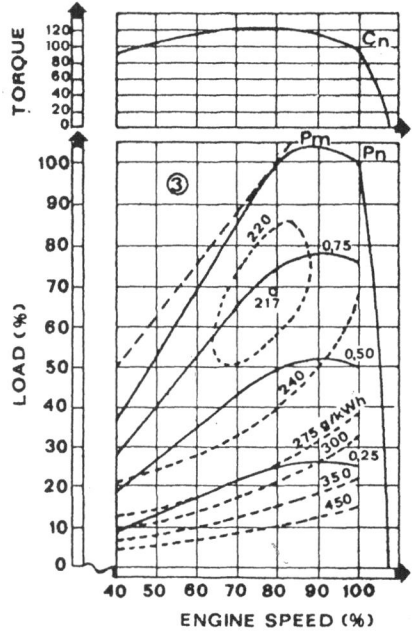

This result can be realized if:

- a large information and professional training program is launched for farmers;

- the manufacturers of farming machines become aware of the real needs of European agriculture and of the necessity to encourage the formation of joint-ventures between the tractor and implement manufacturers, in order to offer optimized full-lines, as the large international manufacturers have already done.

The improved utilization of <u>the working capacity of the implements</u>, is essentially related to the adjustment of the size and shape of the fields for the optimum utilization of the mechanization. It is known that if the fields are small, the working capacity of the implements is reduced, because of the increased incidence of down time on the total working time.

The relationship between down time and field size is illustrated in <u>Fig.63</u>.

<u>Fig.63</u>

<u>Relationship between fields area and down times on total working time of agricultural machines</u>
(Source: Pellizzi, 1986)

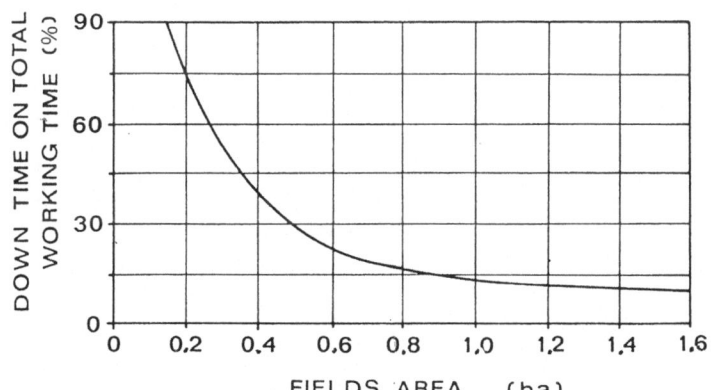

The relationship is described by the following equation:

$$x \cdot y = 0.15$$

A similar correlation exists for the field shape, when the area remains constant. For regular quadrangles, the down times are lower than when the ratio of long side/short side is high and the machines work lenghtwise. The family of curves drawn in **Fig.64** describing this relationship are derived from the equation given below:

$$x^n \cdot y = a$$

where, when a = 1, according to the field size,

$$0.15 \leqslant n \leqslant 0.40$$

Fig.64

Relationship between the down times on total working time, for fields with different side-length ratio
(Source: Pellizzi, 1986)

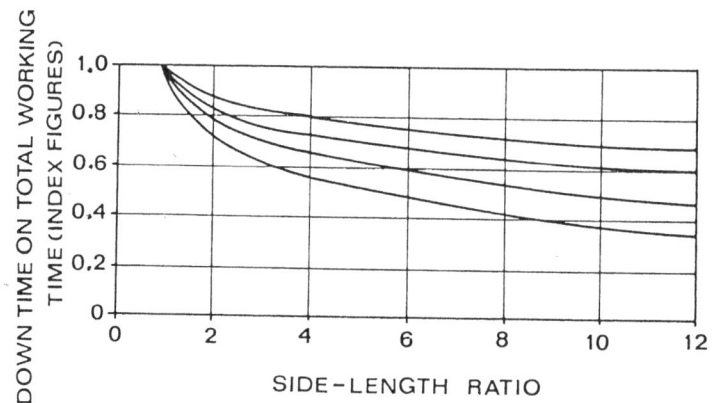

Figures 63 and 64 indicate that there is little advantage in having fields of more than 1.2 ha, while they are significant for smaller fields; for side/length ratios higher than 8:1 the down time reductions are modest.

These considerations are important in respect of the reorganisation costs of the drainage pattern, irrigation network and the arrangement of the farm roads.

Comparisons can also be made between fields shaped as triangles and as rectangles. A triangular field gives rise to 20-30% more down times than a rectangular field having a side-length ratio of 4:1.

If we consider that during dead times the engine runs at only 20-25% of the maximum load, from the curves of Fig.62 the specific fuel consumptions are 40-60% higher than with the maximum load. In fields having dead times of around 40% of the total working time (areas of less than 4000 m2), the fuel consumption of the tractor during dead times is around 35% of the consumption we have during the working time. In fields having dead times of only 10% (areas of more than 1.2 ha) the dead time consumption is only 8% of the consumption for the working time. The difference between the two cases represents a saving of around 12-15% to which must be added the value corresponding to the shorter tractor utilization.

An adaptation of fields to the use of tractors and implements, may result in savings equivalent to that obtained through the optimization of the coupling between engines and implements.

However:

-a large information must be supplied to the farmers;

-incentives must be given to encourage them to meet the necessary costs for the adaptation to the mechanization of the fields and of the other farm structures.

The cultivation systems and farm sizes are different in the various countries. Taking this into account we estimate the energy savings to be about 0.4-0.5 Mtoe/y, for all the EEC.

4.2 - Operative conclusions

It would be worth-while to expand the analysis already developed taking into account the different pedoclimatic regions and the different cultures. The possibilities of reducing energy consumption in this area are clear.

For the EEC countries, the possible savings are conservatively estimated to be 1.3-1.5 Mtoe/y or 15-18 kgoe/ha.y for arable and permanent crops area.

This evaluation has been carried out on the basis of the present energy requirements which are predicted to decrease because of the natural replacement of tractors and machines stock. However, in the three new member countries, where there is a low mechanization level, energy requirements will probably increase.

For the achievement of these objectives it is necessary:

-to put into practice a reliable and general information campaign;

-to create suitable financial incentives for the adaptation of fields to mechanization and for the renewal of existing stock;

-to organize demonstration actions;

-to link the granting of easy term loans to the purchase of certified machines that give optimum performances.

-to supply finance to manufacturers to carry out applied research - in conjunction with public research bodies - to develop new products with lower energy consumptions.

5 - <u>ACTIONS TO DEVELOP AND THEIR IMPACT</u>

5.1 - <u>Summary of the situation and obtainable benefits</u>

The analytical survey -mostly bibliographical- that has been presented leads to the following conclusions:

*-The present direct energy consumptions (fuels and lubricants) for the EEC arable and permanent crop area are around 11-12 Mtoe/y, with an average specific value of 150 kgoe/ha.y.

*-The consumptions can vary by a factor of 5-6 when referred to the specific cultures, to the production intensity, to the various pedoclimatic situations and to the socio-economic levels in the different countries. In other words, the variation range of the single consumptions goes from 60-70 kgoe/ha.y up to 400 kgoe/ha.y.

*-In general, the field operations which contribute most to the global energy consumption are soil tillage + seed-bed preparation (about 40% of the total) and crop harvesting (30-35%).

*-The energy savings which can be attained in the short-medium term depend on:
- the development of the tractor and machine manufacturing industry, through the introduction of innovations that improve the present performances and/or by radical innovations that reduce energy requirements;
- the farmers awareness of the availability of more appropriate machinery, suitable coupling tractor/implements and the optimisation of working schemes;
- the field demonstrations and the dissemination of new techniques (with the same yields) with lower energy consumptions;
- the necessity to adapt the farm structure (fields and equipments) to the optimization of the farming machines;
- the development of jobbing utility enterprises.

*-A global reduction of the present energy requirement by about 30%, that is 3-3.2 Mtoe/y, may be obtained in a few years through well directed and coordinated action. If the improvements already developed were applied, an energy saving of more than 40% could be attained. However it is not realistic to expect a concomitant improvement of both, the efficiency and the management of machines in all farming situations. Therefore, within the next 8-10 years, a consumption reduction of 3-3.2 Mtoe/y, must be considered as an important result.

To publicize and promote energy saving action in the industrial sector appears to be relatively easy, as this sector is concentrated and sensitive to these kinds of opportunities, particularly after the shrinking of the European market in the last 5-6 years. However, it is difficult to influence the farming world in the medium term.

The farmer is not satisfied with reading documentation or watching a promotional film or attending a conference. Before making his choice, he wishes to see the new machines, to test them while working and to examine the potential use on his farm. In other words, the farmer is cautious in his purchases, particularly now, because of the depressed economic situation of European agriculture. The effects of this recession have resulted in a stagnation of the gross salable product, a negative trend in the added value, fewer liquid assets, poor extension of the economic incentives granted by the Governments, and so on.

5.2 - Actions to be continued

What we have outlined so far has been a general view of the situation, confirming the possibility of energy saving in the medium term of up to 30% of the present direct energy requirements. Consequently the production costs would be reduced, in parallel with a technical evolution of: the products; the production processes; the development, use and commercialization of innovations. All these would lead to an increase in saving and, at the same time, to the relaunching of the commercial sectors concerned.

It is necessary to state explicitly the actions required in order to achieve the main objectives.

First we must emphasize that, in order to attain the above mentioned targets, both the industrial and the agricultural sectors will be involved, even though their environmental needs and socio-economic and structural aspects may be extremely different.

An 8-10 year project aiming to reduce the energy requirement of the farming activities without damaging the crops productivity, the farmers revenues and the working conditions of the operators, requires a considerable effort from national Governments to activate their public organisations, research Institutes and technical assistance structures in order to carry out the actions outlined and to be coordinated by the EEC.

To reach these goals, not only the energy aspect of the cultivation operations will be involved, but the whole farming production system will be transformed and renewed, its operators being compelled to manage more complex and sophisticated technologies than at present.

The required actions appear to be extremely complex, but they can be summarized as follows:

a)- Interventions having a politico-informative character. The offices responsible for national agricultural affairs should derive a suitable strategy for the development of structures (in general) and mechanization (in particular);

b)- Interventions to stimulate the farming sector. The aim is to support the structural transformations and to improve the information and training level of the farmers, in order to make it easier for them to understand the current problems and help them making purchasing decisions;

c)- Interventions to assist the industrial manufacturing machines sector (both tractors and implements). To promote applied research and the introducion of technologic developments, particularly improvements in the efficiency of machines and their components;

d)- Setting up, in all the countries, a compulsory machines field performances certification (for both tractors and implements and following standard procedures) as a preliminary condition for granting low interest loans to farmers.

5.2.1 - <u>Interventions in the primary sector</u>

Our analysis shows that the benefits deriving from the techniques and technologies just outlined will only be realized if substantial modifications are introduced in the production structure and if the operators can reach a higher cognitive level.

Therefore, an intervention program in the primary sector must be implemented following two main directions: the first is a <u>structural reform of the sector</u>; the second is to provide technical assistance, and on going extensive and repeated demonstrative actions.

In illustrating the advantages coming from the use of a machine, a technology or a system, we have emphasized that these savings were only permitted by the farm size, the size of the fields and their arrangement, the road network, etc.

The new machines are more complex and have both an increased working capacity and a better efficiency. Often, the investments for these machines appear high and out of place with the structure of the farm. Consequently these machines appear uneconomic: this is especially true for the Mediterranean countries.

A policy that encourages the merging farms, the rationalisation of the structures and of the field sizes, the adjustment of the growing forms in the fruit tree crops, appears to be necessary. However this type of action will be characterized by a strong political component that will make their activation and coordination very difficult.

If properly organized this process will only lead to appreciable results in the long term, i.e. periods longer than the times required for technological developments.

An action carried out on existing structures (to single out in each Region and to transform in a suitable manner) to demonstrate the possible advantages and the positiv benefit/cost ratio, could considerably shorten the necessary time. This action may encourage farmers to pursue, in their farms, the same results shown in the demonstration projects. Financial incentives, in the form of low interest rates loans, would increase the motivation for these transformations.

An action to support the introduction and dissemination of new mechanical agricultural technologies with improved efficiency it may be necessary to promote the formation and the spreading of agro-mechanical utility enterprises. This type of enterprise already has a foothold in the regions where the structures and the farm size are inadequate and in regions having high productivity.

It is necessary to point out that, in order to maximize benefits, the development and use of utility enterprises must be coordinated. Presently these utility enterprises are spreading in response to market demand with unwanted duplication, immoderate competition and loss of efficiency.

Investigations concerning the definition of good development and management models for this type of enterprise, in order to support their establishement and development, could be of great help to the development of new technologies and for improving the efficiency of the whole farming sector.

Actions for the extension, up-dating, and demonstration are implicit in the above assertions. The development lines of these actions could be:

- professional refreshment courses and conference sessions;

- fairs and field demonstration meetings;

- creation of pilot farms for permanent demonstrations;

- machines certification.

With regard to the first point, professional refreshment courses are already numerous in the farming sector due to the availability of EEC grants.

The introduction and dissemination of the above mentioned techniques and technologies requires a series of "ad hoc" courses, prepared on the basis of strict instructions and run by qualified Institutes or Organizations.

It would be desirable to exchange information and experience among the member countries and arrange for the participation of technicians and experts at the same courses in the different countries.

The organization of regular conferences is also considered to be valuable. This proposition appears sensible if we consider how often some companies use this procedure to promote their products, and how successful they are among the agriculture operators.

Considerable value is also assigned to the so called "field fairs", demonstrations which are very popular in the USA but are still not well known in Europe. During this type of demonstration the different implements and systems are illustrated and compared. Together with the permanent pilot demonstration farms, the field fairs seem to provide an irreplaceable opportunity for the farmer to view, handle and learn how the innovative systems may be used on their own farm.

Finally, another valuable instrument is the <u>agricultural machines certification</u>, that is the attestation of the actual characteristics and technical performances of tractors and implements by authorized public Institutions. The farmers would be allowed to evaluate and compare the possible choices. The objective is to make them able to appreciate the performances of the machines and to evaluate the opportunity to buy and introduce them in a particular production cycle.

In this respect the manufacturing aspects of the machines must be considered too. We refer mainly to their life and reliability, and also to the covenience and frequency of the maintenance practices. Often the quality of the agricultural machines (above all of the operating machines) is unsatisfactory, denoting improvisation and insufficient knowledge of the right criteria in the design of various components, leading to a deterioration in the agricultural system efficiency.

In some EEC countries of Northern Europe the machine certification already exists.

In order to maximize the benefits coming from this certification service, it could be useful to bind the granting of financial assistance for the purchase of farm machinery with the exhibition of a certificate that guarantees reduced energy consumption and the efficiency of the mechanized operations. This certificate would only be given where there is a sharp quality improvement in respect to existing performances. It could also provide an indirect motivation for the manufacturing industry to investigate and market energy saving innovations.

However, to avoid discrimination between manufacturers of different size and structure, this procedure would cost the manufacturers an amount in proportion to their budgets.

This means that the cost of this activity could be defined as a function of the market size of the machines requiring a certificate.

The analysis we have carried out on the machines, techniques and technologies shows that some of them are ready to be put into practice; others, are still at the first experimental stage.

To disseminate the advantages already brought out by research, a supplementary effort is necessary, to support applied research.

This action is necessary in the farming sectors which cannot take advantage from the technological systems of other activities. For example, in the field of engines, transmissions, monitoring, automatic control etc., the agricultural developments may originate in sectors completely outside of farming activities. Most of the techniques and technologies specifically relevant to the primary sector have been developed inside the sector itself.

It appears that the following sectors would seem to justify further research activity, finalized and coordinated at the EEC level.

- tyres (for driving or steering wheels) that have less rolling resistance and avoid or reduce soil compaction;
- tractor-implement systems and their optimization (also by CAM methods)
- primary and secondary soil tillage techniques with low energy consumption;
- electric powered systems;
- implements and technologies for pesticide and herbicide distribution having low inputs of chemicals and possible substitution systems;
- machines for (herbaceous and arboreal) harvesting crops and optimization of handling systems.

We have already mentioned the potential savings through improved tyre design. The present research and experimentation efforts are likely to lead to further improvements in respect of the recently developed models.

The research on the second sector is still at an early stage.

In the third sector, the most important consequences of soil tillage are agronomic. The variety of pedoclimatic conditions existing in the EEC countries must be taken into account before we can consider the new techniques useful for most European territories. Research must be carried out in different conditions and the techniques must be tuned to the particular environments. Nevertheless, some technologies, for instance, the diamond mouldboard ploughs and the finger ploughs, could already be used for demonstration.

In spite of the possible benefits of electric propulsion systems, at present their utilization is only occasional and at an early stage. This type of technology is too far ahead of present farming methods and the development lines of manufacturing industry.

To increase the distribution of these innovative farming operations (not only in order to save energy, but also to improve the ergonomic conditions by means the lack of noise and vibrations) it is necessary to support and provide incentives for research, development and demonstration actions.

The same considerations must be made for the new technologies that are able to achieve significant savings in products and other chemical inputs. In this case, the savings are indirect, but the environmental implications are important because they include not only the primary sector but all the community. A coordinated research activity in many disciplines is necessary.

Finally, the research on harvesting systems has to be developed:

-to obtain lower energy inputs,
-to reduce product losses,
-to optimize handling techniques and chains.

5.2.2 - Interventions in the industrial sector

In recent years all the producing countries have recorded (particularly in the EEC countries) a large decrease in agricultural revenue. The decrease in demand for farming machines has induced an economic crisis in the manufacturing industry, with a consequent decline in overall production. As mentioned in § 1, surplus food production is being maintained and, unless new markets (third world) are opened or alternative industrial targets (distillation, fibres, etc.)are found, it is unlikely that manufacturing will return to past production levels. Currently, a further decrease is expected.

In this state of uncertainty manufacturing industry is not inclined to develop technologies and systems which require large investments.

For this reason we believe that external action is needed to motivate the manufacturing industry if we want the new technologies to reach the final users. Legal measures may be useful to stimulate the applied research. However, as mentioned in § 1, the wide distribution of farm machinery production must be kept in mind, and that there are only a few large or medium size producers. Most of the manufacturers are artisans and do not have the capacity to activate their own research lines. In other words, the instruments that theoretically promote production developments, may find an industrial structure which, in practice, is not able to respond.

For these reasons the preliminary actions to promote the manufacturing sector are:

- to encourage the establishment of associations of manufacturers, even though they may be limited to carrying out applied research;

- to promote contacts between the public research Institutes and the aforesaid associations of manufacturers;

- to create information centers that provide technical support for planning and marketing (feasibility studies, profit analysis, market surveys etc.);

- to connect the agricultural machines manufacturing industries and other basic industrial sectors (mechanical, electronic, chemical, etc.);

- to assist manufacturers in the marketing and promotion of their products through certification and field demonstration activities.

In this hypothesis it will be necessary to establish a hierarchy among the different sectors. To do this a detailed analytical study, not within the scope of this present survey, must be carried out.

We believe it would be worth-while to promote the implementation of "full-lines" for the construction of farm machines, taking into consideration the efficiency of the various components. However, this would require an association of the enterprises producing the different components.

With respect to financial incentives it seems opportune to mention the procedure usually followed by farm machinery from the research stage to commercial production. This is a very complex and multi step procedure and in order to actually take advantage of the incentives, it is necessary that all the documentation indicates clearly the possible end-uses, to make it easier to evaluate the usefulness of the proposals. A preliminary analysis can give prominence to the weak and strong aspects of each proposal, so that a preliminary selection can maximize the expected cost/benefit ratio. From this selection it will be necessary to plan, with the collaboration of agromnomic experts, a strategy for the construction of prototypes.

Once the prototypes are manufactured, the following phases should be considered:

- careful testing of the prototypes in various field conditions, and singling out necessary improvements;

- testing a preliminary series of 5-15 units in different agronomic conditions in the marketing regions of most potential: where the results are positive, demonstrations can begin;

- development of the demonstration actions and start up of the certification procedures;

- commercial production and technical assistance in marketing and financial incentives for farmers who need to rearrange their structures if they want to utilize the new tools.

This brief summary list indicates that the research Institutes and extension Services, must be involved together with the pilot plants selected as permanent demonstration centers, in the setting up of the technical training and information service.

5.2.3 - Operative conclusions

The actions suggested by the authors in this document to achieve the targets previously outlined, can be summarized as follows:

- to provide incentives for the modification of farming structures to meet mechanization requirements;

- to support the establishment of agro-technical utility enterprises, that are able to supply a complete service;

- to set up or to strengthen a net of Institutions devoted to the supply of information to farmers and to the dissemination of the innovations by means of refresher courses, meetings, conferences, fairs, field demonstration days and pilot demonstration farms;

- development of a certification service, on a EEC level, for innovative farming machines and implements;

- to promote applied researches to develop immature sectors, linking research Institutes, manufacturing industries and agricultural world;

- to create service centers for farming machines, devoted to supply technical and marketing assistance to small manufacturing enterprises which are not able, because of their size, to participate independantly in the aforesaid innovation procedures.

All the preceding conclusions are valid for field operations but they may also apply to the equipment used for stationary-point operations where, in our opinion, larger energy savings are possible.

Besides this there are opportunities to carry out studies of:

- the actual structure and distribution (in the time, for each operation, for each type of energy) of the direct and indirect energy consumptions in all the Members Countries, with reference to the different crops and to the various pedoclimatic conditions;
- the structure, consistency and problems of the whole manufacturing sector for tractors, agricultural machines and plants, including the induced activities.

These should follow common and standardized methodologies, in order to obtain comparable results.

6 - COST/BENEFIT ANALYSIS AND CONCLUSIONS

6.1 - Preliminary cost/benefit analysis

As we have already outlined, the energy savings obtainable, in a steady regimen, by applying and disseminating new systems and technologies can be estimated at around 3 Mtoe/y that is, roughly, around 0.7-0.8 billion ECU/y. This figure may not be considered relevant if compared with the Gross Salable Product (GSP) of EEC agriculture but it is sgnificant if it is referred to the field vegetable production, which represents half of the whole GSP. The figure is also important if compared to the annual spendings on agricultural machines and implements.

We must bear in mind that if figures mentioned above represent the possible saving in direct expenditures, then the techniques and machines that produce this saving, will at the same time induce other indirect economic benefits, for example:

- a higher quality of work, mainly from the agronomic sector;

- an improved global efficiency, with substantial indirect savings, particularly in the agrochemical products used;

- lower production losses: which corresponds to an increase of production itself;

- greater respect for the environment (less pollution, less erosion etc.) and towards man;

- less energy content in the newly developed machines.

The economic benefits from the new technologies are therefore likely to be much higher than 0.7-0.8 billions ECU/y. Even though it is difficult to evaluate, the global indirect advantages of well directed research, demonstration, popularization etc., should result in a potential saving of three times that already given. To the total must be added the benefits from the relaunching of the industrial sectors.

6.2 - Conclusions

The analysis we have carried out has been essentially bibliographical, without any direct experimental analysis. Nevertheless, it has been confirmed that there is considerable scope for incremental and radical innovations in European agricultural mechanization, resulting in significant energy savings and improved global efficiency.

This analysis has also been an opportunity to show how

European manufacturers are sensitive to rationalisation problems. This has occurred after a period where overseas technologies were used which were not suited to European agriculture. This trend was not welcomed by the European farmers who, in the economic crisis, have been more careful and attentive than previously in their decision-making and investments.

The analysis has also outlined:

-    the validity of the European activity in this sector which should be improved, whether at EEC or national level;

-    the opportunity to develop the same type of analysis in fields, like: the energy consumption structure; the possibilities of energy savings in the stationary farming operations, the indirect energy savings in the chemical inputs; the structure and distribution of the manufacturing sector.

-    the opportunities (and the possible benefits) coming out from a series of coordinated actions in R&D and demonstration in favour of the agricultural and industrial sectors, having as a target the rationalization of energy consumption in the two sectors.

7 - MAIN REFERENCES

7.1 - National energy balance and energy saving in general

1)--Bartorelli M., Pellizzi G. et Al.: "Il sistema agro-silvo alimentare in rapporto all'energia" - CNR, n.24, 1982.

2)--Carillon R.: "Essai sur l'énergie dans l'agriculture et dans le système agro-alimentaire en France" - Etude du CNEEMA, n.404, 1975.

3)--Carillon R.: "La consommation d'énergie dans l'agriculture" - Bulletin d'information du CNEEMA, n.217, 1976.

4)--Carillon R.: "L'activité agricole et l'énergie" - Etude du CNEEMA, n.408, 1978.

5)--Carillon R.: "L'analyse énergetique de l'acte agricole" - Etude du CNEEMA, n.458, 1979.

6)--Carillon R.: "Le bilan énergetique de l'agriculture française et les conclusions qu'on en dévrait tirer" - Cahiers du CENECA, 122, 1980.

7)--Carillon R.: "L'exploitation agricole et l'énergie: les economies d'énergie" - B.I. du CEMAGREF, n.292, 1982.

8)--Carillon R.: "Confirmation de la consommation d'énergie directe de l'agriculture" - B.I. du CEMAGREF, n.322, 1984.

9)--Carillon R.: "La consommation d'énergie directe dans les exploitation agricoles" - B.I. du CEMAGREF, n.331-332, 1985.

10)-CEMAGREF: "Contribution à la saisie de la consommation d'énergie en agriculture" - Etude n.505, 1983.

11)-CNEL: "Osservazioni e proposte su agricoltura ed energia" - CNEL, Roma, 1982.

12)-Coolman F., de Vries R.L.: "Tendences actuelles et prévisibles de la mechanisation de l'agricolture (horizon 1990)" - Rapport AGRI/MECH., U.N. Genève, n.92, 1981.

13)-Dufey V., Misson A.: "Economie d'énergie en mechanisation agricole" - Station de Genie Rural, Gembloux, Agro-Service 2, 1982.

14)-**: "FAO production yearbook" - FAO, 1984.

15)-FAO: "Report of the 1984 technical consultation of the european cooperative network on rural energy: energy conservation with tractors and agricultural machines" - CNRE, 4, 1984.

16)-Herview J.F.: "Délibération du conseil supérieur de la mécanisation et de la motorisation de l'agriculture sur les problèmes d'énergie" - B.I. du CNEEMA, n.259-260, 1979.

17)-Kastroll H.J.: "Aktuelle Probleme der Energie Produktion und Energie Anwendung in der Landwirtschaft" - 10th Int. Cong. of Agric. Eng., 85-139, sept.1984.

18)-Leach G.: "Energy and food production" - Science and Technology Press, 1976.

19)-Leach G.: "Energy and food production" - Energy Analysis, Ed. J. Thomas, Westriew Press, 1977.

20)-Lebailly J.P.: "Le bilan énergétique de l'agriculture belge: deux instants 1959-61 et 1974-76" - Cahiers du CENECA, n.129, 1980.

21)-Matthews J.: "The power requirement for tillage in the next decade" - The Agriculture Engineer, 99-104, Winter 1979.

22)-Mellade L., Nadal-Amat A.: "Programme d'agrénergétique I.N.I.A. (Espana)" - Cahiers du CENECA, 2219, 1980.

23)-Mercier J.R.: "Energia e agricoltura" - Muzzio Ed., 1980.

24)-Ministerio de Agricultura, Pesca y Alimentacion: "Report on agricultural statistics" - Madrid, 1983.

25)-Moens A.: "Energy consumptions of agricultural operations" - Riv. d'Ingegneria Agraria, 140-144, n.4, 1977.

26)-Nielsen H.: "Besoins en énergie de l'agriculture danoise et rôle possible de celle-ci pour la production d'énergies de substitution" - Cahiers du CENECA, 2210, 1980.

27)-Novacki T.K., Novacki J.K.: "Evolution et tendances du volume et du bilan de la consommation d'énergie dans l'agriculture" - Rapport AGRI/MECH., U.N. Genève, n.105, 1984.

28)-OECD: "Le secteur agro-alimentaire face au problème de l'énergie" - Rapport général, 1981.

29)-Oksanen E.H.: "The significance of energy conservation in mechanized farming" - CNRE, n.4, FAO, 1984.

30)-Pellizzi et Al.: "Agricoltura e crisi energetica" - CNR, n.18, 1981.

31)-Pernkopf J.: "Energy conservation with tractors and agricultural machinery" - CNRE, n.4, FAO, 1984.

32)-Salooja K.C.: "Energy conservation in farming" - ESSO Research Centre, 1979.

33)-Severkev, Tsyganov: "Economie d'énergie dans les travaux de culture en Europe" - B.I. du CIMAGREF, n.292, 1982.

34)-Slesser M.: "Energy use in the food-production sector of the European Economic Community" - Energy and Agriculture, Springer-Verlag, 1984.

35)-Therasse R.: "Le remembrement et la réduction de la consommation de carburant" - Agricontact n.113, février 1981.

36)-Thoma H., Ortmaier E., Stürmer H.: "Auswirkungen steigender Energiepreise auf die Kosten variabler Input-Komponenten in der landwirtschaftlichen Produktion der BR Deutschland" - 1st Int. Conf. "Energia e Agricoltura", Milano, 1983.

37)-Toniolo L.: "Tecniche agronomiche e loro riflessi sul bilancio energetico" - Rivista d'Ingegneria Agraria, 151-160, n.4, 1977.

38)-UNACOMA: "L'industria italiana delle macchine agricole"

- Roma, 1983.

39)-Van Hecke E.: "A regional approach to the analysis of Belgian agricultural energy use and production" - Energy in Agriculture, 117-130, 1980-1982.

40)-Weber A.: "Bewertung der Energiebilanz aus Produktion, Distribution und Verbrauch" - Agrarwirtschaft und Energie, Verlag Paul Parey, 98-113, 1979.

41)-Wenner H.: "Energiebedarf und Energieanfall in der Landwirtschaft" - Landtechnik, 6-10, n.1, 1983.

7.2 - Crops and agricultural practices energy requirements

1)--Baldini et Al.: "Analisi energetiche di alcune culture arboree da frutta" - Riv. d'Ingegneria Agraria, 73-201, n.2, 1982.

2)--Baldi F., Spugnoli P.: "Computo dei flussi energetici dovuti alle macchine agricole. Applicazione al caso di una azienda viticola e di una olivicola della collina toscana" - Istituto di Meccanica Agraria, Firenze, 1985.

3)--Biondi P., Bolli P., Farina G., Panaro V.: "Analisi dei consumi e delle possibilità di risparmio energetico in una azienda cerealicolo-zootecnica" - Macchine e Motori Agricoli, n.8, 1982.

4)--Cavalli R.: "Prime valutazioni di moderne tecnologie per una integrale meccanizzazione delle operazioni colturali della patata" -Atti Conv. Naz. di Meccanica Agraria, Perugia, 1985.

5)--Cavazza L. et Al.: "Analisi del consumo energetico della barbabietola da zucchero in Italia" - Riv. di Agronomia, 213-228, 1983.

6)--Dohne E.: "Energie Fragen in der Landwirtschaft" - Landtechnik, 457-460, n.11, 1977.

7)--Dwyer M. J.: "Power requirements for field machines" - Agricultural Engineer, 50-59, Summer 1985.

8)--EEC: "Energy consumption per tonne of competing agricultural products available to EC" - Information on Agriculture, n.85, 1981.

9)--EEC: "Analyse comparative des structures agricoles au niveau régional de l'Espagne, de la France, de la Grèce, de l'Italie et du Portugal devant les perspectives de l'elargissement de la CEE" - Information sur l'agriculture, n. 87 et 88, 1983.

10)-Finassi A.: "Alcune proposte per la riduzione del consumo energetico nella risicoltura italiana" - Rivista d'Ingegneria Agraria, 193-195, n.4, 1977.

11)-Guidobono Cavalchini A.: "Valutazione delle diverse tecniche di raccolta e conservazione dei foraggi prativi in funzione del rendimento energetico" Riv. d'Ingegneria Agraria, 179-192, n.4, 1977.

12)-Guidobono Cavalchini A., Natalicchio E.: "Costi energetici del riso e del mais, in diversi sistemi produttivi della pianura padana" - 1a Conf. Intern.

"Energia e Agricoltura", 1983.

13)-Guidobono Cavalchini A. et Al.: "Sottoprodotti della barbabietola da zucchero: raccolta a fini zootecnici" - L'Informatore Agrario, 29-36, n.22, 1984.

14)-Heyland K.U., Solansky S.: "Energieeinsatz und Energieumsetzung im Bereich der Pflanzenproduktion" - Agrarwirtschaft unt Energie, Verlag Paul Parey, 1979.

15)-Hollmann P.: "Struktur der Energieeinsatzes und der Energiekosten in Betriebsgrössen und Betriebsformen" - Agrarwirtschaft und Energie, Verlag Paul Parey, 1979.

16)-Hornacek M.: "Application de l'analyse énergétique à 14 exploitations agricoles" - Etudes du CNEEMA, n.452, 1979.

17)-Hutter W., Rouman I.: "Energie consommée par la production de quelques cultures" - Comptes-rendus Académie d'Agriculture de France, n.4, 1976.

18)-Hutter W.: "Efficience de l'énergie consommée par les travaux culturaux" - Bulletin Technique d'Information, n.330, 1978.

19)-Hutter W.: "Reflexions sur le bilan énergétique des cultures" - Cahiers du CENECA, n.121, 1980.

20)-O' Callaghan J.R.: "Energy use in agriculture" - Riv. d'Ingegneria Agraria, 145-150, 4, 1977.

21)-Pick E.: "Rationalisation of energy demand in field operations" - CNRE, Bull.4, FAO, 1984.

22)-Pimentel D.: "Handbook of energy utilization in agriculture" - CRC Press, 1980.

23)-Planeta A.: "Ricerca sul fabbisogno di energia per la produzione del grano duro in Sicilia" - Riv. d'Ingegneria Agraria, 177-196, n.3, 1975.

24)-Prache J.L.: "Elements d'appréciation de quelques intrants énergétiques en grande culture" - B.I. du CNEEMA, 259-260, 1979.

25)-Spugnoli P.,Zoli M.: "Rendimento energetico di una azienda zootecnica con impianto di digestione anaerobica" - Atti Convegno di Meccanica Agraria, Bari, 1981.

26)-White D.J.: "Energy in agricultural system" - The Agricultural Engineer, 52-58, Autumn 1975.

7.3 - Energy saving through a more appropriate use of agricultural machines

1)--: "Economie de carburant dans l'utilisation des tracteurs agricoles" - Agricontact n° 105, Juin 1980.

2)--AFME: "Consommation de carburant des tracteurs agricoles Féviers 1986" - n°1, Paris, 1986.

3)--Audsley E., Dumont S., Boyce S.: "An economic comparation of methods of cultivating and planting cereals, sugar beet and potatoes and their interaction with harvesting, timeliness, and available labour by linear programming" - J. Agric. Eng. Res., 283-300, n.23, 1978.

4)--Bryan Davies D.: "Cultivation systems in the '80" - The Agricultural Engineering, 94-95, Winter 1979.

5)--Cordier Y., Frankinet M., Vitlox O.: "Labour, demi-labour ou semis direct en continu. Conséquences géotechniques" - Bull. Rech. Agronomiques, Gembloux, 2, 1979.

6)--Detraux F.: "La métrologie moderne appliquée aux essais de tracteurs" - Agricontact, n.168, décembre 1985.

7)--Dufey V.: "Mieux connaître le moteur du tracteur pour en tirer le meilleur parti" - Agricontact, n.139, avril 1983.

8)--Dufey V., Pletinckx A.: "Economie d'énergie et mécanisation agricole" - 1st Int. Conf. "Energia e Agricoltura", Milano, 1983.

9)--Dufey V.: "Economie d'énergie et mécanisation agricole" - CNRE, n.4, FAO, 1984.

10)-Dubois J.P., Godon C.: "Tracteurs et economie d'énergie" - Agricontact, n.137, février 1983.

11)-Dwyer M.J.: "Maximizing agricultural tractor performance by matching weight, tyre size and speed to the power available" - Proc. 6th Int. ISTVS Conference, Vienna, 1978.

12)-Gego A.: "Measurements on tractors design for reducing energy requirements" - Riv. d'Ingegneria Agraria, 234-238, n.4, 1977.

13)-Hernanz J.L. et Al.: "Ensayos de laboreo minimo y siembra directa en la Zona Centro" - 2° Jornadas Técnicas sobre cereales de invierno, Pamplona, 1985.

14)-Janin J.L.: "Moyens de contrôle et gestion optimale pour le choix et l'utilisation des tracteurs" - CNRE, n.4, FAO, 1984.

15)-Janin J.L.: "Moyens de contrôle et gestion optimale pour le choix et l'utilisation des tracteurs" - 10th Int. Congr. of Agricultural Engineering (CIGR), 435-442, 1984.

16)-Matthews J.: "Effective choice and use of agricultural tractors" - The Agr. Engineer, 96-100, Autumn 1982.

17)-Motamoros M.E., Enebral M.M.: "Elements techniques contribuant à l'utilisation optimum des machines agricoles" - Rapport AGRI/MECH, U.N., Genève, n. 104, 1984.

18)-Nielsen V.: "Undersogelser vedr. reduceret jordbehandling" - Statens Jord. Forsog, Beretning n.14, May 1982.

19)-Nielsen V.: "Energioforbruget ved händtering og behandling af groufoder" - Statens Jord. Forsog, Orientering n.40, August 1985.

20)-Norén O.: "Different methods of forage harvesting from an energy point of view" - CNRE n.4, FAO, 1984.

21)-O'Dogherty M.J.: "A review of research on storage chopping" -J. Agr. Eng. Res., 267-289, 1982.

22)-Oestges O.: "Puissance requise par les machines

agricoles" - Agricontact, n.139, avril 1983.

23)-Pellizzi G.: "On the rationalization of agricultural mechanization in relation to structural reforms" - Riv. d4Ingegneria Agraria, n.1, 1976.

24)-Perdok U.D., Lamers J.G.: "Studies of controlled agricultural traffic in the Netherlands" - Int. Conference of Soil Dynamics, June 1985.

25)-Petterson D.E., Chamen W.C.T., Richardson C.D.: "Long-term experiments with tillage systems to improve the economy of cultivations for cereals" - J. Agric. Eng. Res., 1-35, 1980.

26)-Petterson D.E.: "The performance of alternative cultivation systems" - Agricultural Engineer, 8-11, Spring 1982.

27)-Piccarolo P.: "Meccanizzazione razionale e organizzazione del lavoro per ridurre le spese di energia in agricoltura" - Riv. d'Ingegneria Agraria, 161-172, n.4, 1977.

28)-Piel-Desruisseau J.: "L'organisation du travail en agriculture" - Ed. Organisations, Paris, 1963.

29)-Reich R.: "Aspetti tecnici e pratici sull'utilizzazione del coltivatore nella lavorazione delle stoppie e del terreno" - Macchine e Motori Agricoli, 39-42, n.7, 1982.

30)-Steinkampf H.: "Energiesparung in der Pflanzenproduktion-Bereich Agrartechnik Kampf" - Agrarwirtschaft und Energie, Verlag Paul Parey, 157-167, 1979.

31)-Steinkampf H.: "Wirtschaftlicher Einsatz leistungsstarker Schlepper aus energetischer Sicht" - Landbauforschung Völkenrode, n.27, 159-164, 1977.

32)-Steinkampf H., Von Gerhard J.: "Einflussgrössen auf Flächeleistung und Energieaufwand beim Schleppereinsatz" - Grundl. Landtechnik, n.1, 20-27, 1982.

33)-Steinkamer H., Jahns G.: "Optimierung der Zuordnung von Schlepper und Gerät" - 10th Int. Congr. of Agricultural Engineering (CIGR), 443-450, 1984.

34)-Stroppel A.: "Possibilità di risparmio energetico nella lavorazione del terreno" - Macchine e Motori Agricoli, 29-34, n.7, 1982.

35)-Trolli C.: "Risparmio energetico nell'impiego dei pneumatici in agricoltura" - Macchine e Motori Agricoli, 21-26, n.10, 1983.

36)-Val Manterola et Al.: "Medida de los consumos énergeticos de un tractor empleado para las operaciones mecanizadas del cultivo de los agridos" - 17a Conferencia Intern. de Mecanizacion Agraria, Zaragoza, 1983.

37)-Vitlox O.: "Economie d'énergie en labour" - Agricontact, n.141, juin 1983.

38)-Vitlox O.: "Economies d'énergie par l'appairement du tracteur et des outils" - CNRE, n.4, FAO, 1984.

39)-Witney B.D., Oskovi E.K.: "The basis of tractor power

selection on arable farms" - J. Agr. Eng. Res., 513-527, 1982.

40)-Woerman G., Bashford L.: "How much does front wheel assist really help?" - Agricultural Engineering, 31-35, n.4, 1984.

7.4 - Energy saving due to product and process innovations

1)--: "Pneumatiques agricoles et économie d'énergie" - Agricontact, n.115, avril 1981.

2)--: "Bois ou paille pour un moteur: 80% de fuel économisé" - Motorisation et Technique Agricole, 63-64, Octobre 1981.

3)--: "Farm Electronics and Computing - International Symposium" - Royal Agricultural Society of England, London, 1985.

4)--: "Nachwachsende Rohstoffe und Energietechnik" - Arbeiten des Instituts für landtechnische Grundlagenforschung, Braunschweig-Völkenrode, 1980-1985.

5)--: "Mikroelektronik für die Agrarproduktion" - Arbeiten des Instituts für landtechnische Grundlagenforschung, Braunschweig-Völkenrode, 1980-1985.

6)--Abeels P.F.: "Pont auxiliaire moteur; contribution à l'économie en énergie" - Unité de Génie Rural, Univ. Cath. de Louvain, 1980.

7)--Abeels P.F.: "Rim, tyre and soil" - Unité de Génie Rural, Univ. Cath. de Louvain, 1979.

8)--Abeels P.F.: "L'amélioration de la traction mécanique à la ferme" - CNRE, n.4, FAO, 1984.

9)--Achart J.: "Possibilité de réduction de la consommation d'énergie des moteurs et des tracteurs au niveau europeen" - B.I. du CEMAGREF, n. 298, 1982.

10)-Achart M.: "Evolution du moteur Diesel en agriculture" - Motorisation et technique agricole, 73-77, mai 1984.

11)-Alcock R.: "Battery powered vehicles for field work" - Transactions of the ASAE, 10-13, 1983.

12)-Baldini E. et Al.: "Possibilità di sviluppo della produzione di macchine per la raccolta e la potatura delle produzioni arboree e per la raccolta delle produzioni ortive" - CESMA, Reggio Emilia, giugno 1985.

13)-Bassi A.: "Prospettive del motore automobilistico in relazione all'ambiente" - Acqua e Aria, 31-38, n.1, 1986.

14)-Baskin G.R., Mayeux M.M., Sistler F.E.: "A monitor for crop moisture and cylinder speeds" - Transactions of the ASAE, 1220-1224, 1982.

15)-Biller R.H.: "Automatische Datanerfassung beim Schlepper Einsatz" - Int. Congr. of Agricultural Engineering (CIGR), 249-256, 1984.

16)-Blight D.P.: "Research into powered cultivations" - The Agricultural Engineer, 96-98, winter 1979.

17)-Bodria L., Guidobono Cavalchini A., Lazzari M.: "La raccolta dei cereali in pianta intera" - L'Informatore Agrario, 51-58, n.2, 1984.

18)-Brennedörfer M.: "Automatisation de la commande des processus sur les machines agricoles mobiles" - B.I. du CEMAGREF, n.294, 1982.

19)-Brizgis L.J., Nave W.R., Paulsen M.R.: "Automatic cylinder-speed control for combines" - Transactions of the ASAE, 1066-1071, 1980.

20)-Cera M., Cavalli R.: "Innovazioni tecnologiche nel settore delle lavorazioni del terreno" - Riv. d'Ingegneria Agraria, n.3, 1984.

21)-Chancellor W.J., Thai N.C.: "Automatic control of tractor transmission ratio and engine speed" - Transaction of the ASAE, 642-646, 1984.

22)-Cox S.W.R.: "Microelectronics in agriculture and horticulture" - Granada Publishing, 1982.

23)-D'Addea N.: "Innovazioni nei cicli di produzione" - Rivista d'Ingegneria Agraria, n.3, 1984.

24)-De Zanche C.: "Possibilità di impiego del biogas nella trazione agricola" - L'Informatore Agrario, n.38, 1982.

25)-Dufey V.: "Pour une boîte de vitesses à la prise de force" - Agricontact, n.110, novembre 1980.

26)-Duosi G.G.: "Innovazioni di prodotto nelle trattrici agricole" - Riv. d'Ingegneria Agraria, n.3, 1984.

27)-Eimer M.: "Entwicklungen von Regelungseinrichtungen am Mähdrescher" - Grundlangen der Landentechnik, 137-140, n.5, 1970.

28)-Elsbett L.: "New diesel development for the use of vegetable oil as fuel" - Elsbett Konstruktion, Hilpoltstein.

29)-Fankhauser J. et Al.: "Erfahrungen mit Biogas als Treibstoff für Landwirtschafttraktoren" - FAT, n.27, 1985.

30)-Fekete A., Földesi I.: "Fuel saving by automatic control of tractors and combine harvesting" - CNRE, n.4, FAO, 1984.

31)-Gasparetto E., Trolli C., Viola L.: "Economie d'énergie directe par l'emploi de différent pneus pour tracteurs" - CNRE, n.4, FAO, 1984.

32)-Gasparetto E., Trolli C., Viola L.: "Proposta di impiego del microprocessore per il risparmio energetico nell'uso del trattore" - Riv. d'Ingegneria Agraria, n.3, 1984.

33)-Gervasio V. et Al.: "Prototipo di trattrice alimentata a batteria" - Riv. d'Ingegneria Agraria, n.3, 1984.

34)-Goupillon J.F.: "Production d'énergie mécanique en petite et moyenne puissance par gazeification de la biomasse: avis et travaux du CEMAGREF" - B.I. du CEMAGRF, 307-308, 1983.

35)-Grevis-James I.W., Bloome P.D.: "A tractor power

monitor" - Transaction of the ASAE, 595-597,1982.

36)-Guidobono Cavalchini A., Natalicchio E.: "Per una meccanizzazione più appropriata ed un sistema di produzione adeguato alle aree marginali" - Riv. d'Ingegneria Agraria, n.3, 1984.

37)-Guidobono Cavalchini A.: "Nuove macchine per la preparazione del terreno: prove comparative tra un nuovo tipo di erpice trainato ed un erpice rotativo" - L'Informatore Agrario, 17257-17264, n.37, 1981.

38)-Guidobono Cavalchini A., Lazzari M., Pergher G.: "Indagine sul funzionamento di mietitrebbiatrici nella Pianura Padana" - Ingegneria Agraria, 160-167, n.3, 1985.

39)-Hooper A.W., Ambler B.: "A combine harvester discharge meter" - J. Agr. Eng. Res., 1-10, n.24, 1979.

40)-Huisman W.:"Design method and simulation results obtained with micro processor-based speed-control systems in minimizing the costs of cereal combine harvesting" - 10th Int. Congr. of Agricultural Engineering (CIGR), 265-273, 1984.

41)-Jenny E.: "Progress with medium and low speed engines" - High Speed Diesel Report, 62-68, August 1985.

42)-Kanetoh Y.: "Driveless combine harvester" -Grain and forage harvesting, ASAE, 1978.

43)-Kihara R.: "Progress with high speed engines" - High Speed Diesel Report, 142-145, August 1985.

44)-Kirk T.G. et Al.: "Evaluation of a simulation model of the combine harvester" - Grain and Forage harvesting, ASAE, 1978.

45)-Kruse J., Krutz G.W., Auggins L.F.: "Computer controls for the combine" - Agricultural Engineering, 7-9, February 1983.

46)-Kutzbach H.D.: "Dresch-und Trenn-systeme neuer Mähdresher" - Landtechnik, 226-230, n.6, 1983.

47)-Laib L., Komandi G.: "Measurement of the output performance - History of agricultural tractors" - 10th Int. Congr. of Agricultural Engineering (CIGR), 516-525, 1984.

48)-List H.: "Diesel development in perspective" - High Speed Diesel Report, 50-54, August 1985.

49)-Manby T.C.D.: "Future trends in the development of tractor design to reduce energy input; tractor and implement development and research to improve efficiency". - Riv. d'Ingegneria Agraria, 221-230, n.4, 1977.

50)-Matthews J.: "The mechanical farm of 2030" - Agricultural Engineer, 30-32, Spring 1982.

51)-Mc Nulty D.B., Khaldoon Al-Jobouri: "Potential for energy conservation in vibratory potato digging" - Agricultural and Food Engineering Dept., Dublin, 1984.

52)-Metamoros E., Enebral M.: "Elements téchnologiques pour l'utilisation optimale des materiels agricoles" - B.I. du CEMAGREF, n.329, 1985.

53)-Metamoros E., Enebral M.: "Elements téchniques

contribuant à l'utilisation optimale des machines agricoles" - B.I. du CEMAGREF, n.335, 1985.

54)-Millar G.H.: "Future trends in the development of tractor design in order to reduce energy imput" - Riv. d'Ingegneria Agraria, 231-233, n.4, 1977.

55)-Millar G.H.: "Diesel developments in perspective" - High Speed Diesel Report, 44-50, August 1985.

56)-Molle J.F.: "L'ethanol-carburant" - B.I. du CIMAGREF, n.334, 1985.

57)-Moller A.: "Application of microprocessors within agriculture" - Danish Council of Technology Project, n.1982-133/001-82.035.

58)-Palmer J.: "Electronics and field machinery" - The Agricultural Engineer, 42-43, Summer 1980.

59)-Pang S.N., Zoerb G.C., Wang G.: "Tractor monitor based on indirect fuel measurement" - Transactions ot the ASAE, 994-998, 1985.

60)-Pellizzi G., Piccarolo P.: "Prospettive di produzione di full-lines di macchine per un rilancio dell'agricoltura mediterranea" - Atti Conv. Macchine Agricole e Artigianato, Bari, n.9, 1979.

61)-Pellizzi G.: "Tecnologie meccaniche ed informatiche per l'agricoltura dopo l'età dello spreco" - Atti Conv. Naz. Federconsorzi, Piacenza, 10, 1982.

62)-Pellizzi G.: "La meccanizzazione per l'agricoltura italiana degli anni '90" - Conv. "Meccanizzazione Agricola anni '90", Bologna, n.11, 1983.

63)-Pellizzi G.: "Problemi ed esigenze di innovazione tecnologica nelle macchine e negli impianti per l'irrigazione" - Macchine e Motori Agricoli, n.9, 1985.

64)-Pellizzi G.: "Meccanica e meccanizzazione agricola" - Edagricole, Bologna (in corso di stampa).

65)-Pernkopf J.: "L'emploi d'huile vegétale dans les moteurs diesel: une garantie en cas de crise" - Cahiers du SENECA, n.326, 1980.

66)-Peyrichou B.: "Electronique et informatique en agriculture" - B.I. du CEMAGREF, 307-308, n.29-34, 1983.

67)-Renaud J.: "L'électronique embarquée sur les matériels agricoles" - Motorisation et techniques agricoles, n. 3-4-5, 1984.

68)-Renaud J.: "Mariage de raison: Diesel et gazo-bois" - Motorisation et Technique Agricole, 33-37, n.25, 1981.

69)-Richardson N.A. et Al.: "Getting a true measure of ground speed" - Agricultural Engineering, 14-20, June 1984.

70)-Severner M.M., Lartchenco L.V.: "Amélioration des machines d'épandage d'engrais chimiques pour assurer un épandage plus uniforme sur les terres agricoles" - Rapport AGRI/MECH., U.N. Genève, n.103, 1984.

71)-Shoup W.D., Macchio V.R.: "Agricultural robots: their promise and potential" - Agricultural Engineering, 25-30, April 1984.

72)-Shueller J.K., Mailander M.P.: "Combine feed rate sensors" - Transactions of the ASAE, 2-5, 1985.

73)-Smith L.A.: "Controlling engine speed precisely" - Agricultural Engineering, 9-10, March 1985.

74)-Stadler E., Studer R.: "Untersuchung über den Betrieb von Landwirtschaftstraktoren mit Dieselholzgas" -FAT, n.12, 1981.

75)-Stange K. et Al.: "Microcomputer goes to the field to gather tractor test data" - Agricultural Engineering, 21-25, January 1984.

76)-Tampkins F.D., Wilhelm L.R.: "Micro computer-based tractor data acquisition system" - Trans. of the ASAE, 1540-1543, 1982.

77)-Terrier H.: "Dodge roulant au végétal" - Motorisation et Technique Agricole, 29-31, n.26, 1981.

78)-Vaing G.: "Conception, réalisation et essais d'un prototype de tracteur à gazogène de petite puissance destiné aux pays tropicaux" - Machinisme Agricole Tropical, 3-43, n.81, 1983.

79)-Watts C.W., Patterson D.E.: "The development and assessment of high speed shallow cultivation equipment for autumn cereals" - J. Agric. Eng. Res., 115-122, 1984.

80)-Withers J.: "The past, present and future of the agricultural tractor" - Agricultural Engineer, 74-79, Autumn 1983.

81)-Young S.C., Johnson G.E., Schafer R.L.: "A vehicle guidance controller" - Trans. of the ASAE, 1340-1345, 1983.

82)-Zoerb G.C., Pang S.: "A new method of measuring tractor fuel efficiency during field operations" - 10th Int. Congr. of Agricultural Engineering (CIGR), 202-209, 1984.

8 - RESEARCH INSTITUTES INVOLVED IN ENERGY SAVING IN MECHANIZATION

Belgium & Luxembourg

--Station de Génie Rural, Gembloux
--Unité de Thermodynamique de l'Université Catholique - Louvain la Neuve
--Faculté de Sciences Appliquées - Université Libre de Bruxelles
--Rijksuniversiteit - Gent
--Ecole Royale Militaire - Bruxelles

Denmark

--Statens Jordbrugsteknisk Forsog - Horsens

France

--CEMAGREF - Antony

Federal Republic of Germany

--Institut für Landtechnische Grundlagenforschung der FAL - Braunschweig.
--Institut für Landtechnik - Liebig Universität - Giessen
--Institut für Maschinenkonstruktion, Landtechnik und Baumaschinen - Technische Universität - Berlin
--Institut für Agrartechnik - Universität Hohenheim - Stuttgart
--Institut für Landmaschinen - Technische Universität - Braunschweig
--Institut für Landmaschinen - Technische Universität - München
--Institut für Landtechnik - Technische Universität München-Freising
--Institut für Betriebstechnik der FAL - Braunschweig

Greece

--Institute of Agricultural Engineering - University - Athens
--Institute of Agricultural Engineering - Democritos University - Tessaloniki

Ireland

--Agriculture Institute - Research Centre - Dublin
--University College - Dept. of Agricultural Engineering -
Dublin

Italy

--Istituto di Ingegneria Agraria - Università - Milano
--Istituto per la Meccanizzazione Agricola -CNR - Torino
--Istituto di Meccanica Agraria - Università - Padova
--Istituto di Meccanica Agraria - Università - Udine
--Istituto di Meccanica Agraria - Università - Firenze
--Istituto di Meccanica Agraria - Università - Bari
--Istituto di Meccanica Agraria - Università - Catania

The Netherlands

--IMAG - Wageningen
--Institute of Agricultural Engineering - University -
Wageningen

Portugal

--Instituto Nacional de Investigaçao Agraria e de Extensào
Rural - Oeiras

Spain

--Escuela Tecnica Superior dos Ingenieros Agronomos -
Madrid

United Kingdom

--National Institute of Agricultural Engineering - Silsoe,
Bedford
--Scottish Institute of Agricultural Engineering - Bush
Estate, Midlothian
--Institute of Agricultural Engineering - University - New
Castle upon Tyne

9 - NATIONAL MANUFACTURERS ASSOCIATIONS

Belgium

--Fédération des Entreprises de l'Industrie des Fabrications Métalliques (FABRIMETAL)
--Groupe "Matériels pour l'Agricolture, l'Horticulture et l'Elevage" - Bruxelles

Denmark

--Danske Landbrugmaskinfabrikanter - Foreningen AF 1983 - Aarhus

France

--Syndicat général des constructeurs de tracteurs et machines agricoles (SIGMA) - Paris

Federal Republic of Germany

--Landmaschinen und Ackerschlepper-vereinigung im VDMA - Frankfurt/Main - Niederrad

Italy

--Unione Nazionale Costruttori Macchine Agricole (UNACOMA) - Roma

The Netherlands

--Nederlandse Vereniging van Fabrikanten van Landbouwwerktuigen (N.V.F.L.) - Zoetermeer

Spain

--Groupement National de Fabricants de Machines Agricoles (ANFAMA) - Madrid

United Kingdom

--The Agricultural Engineers Association Limited (A.E.A.) - London